MATERIALS AND MANUFACTURING TECHNOLOGY

METAL MATRIX COMPOSITES

MATERIALS AND MANUFACTURING TECHNOLOGY

Additional books in this series can be found on Nova's website under the Series tab.

Additional E-books in this series can be found on Nova's website under the E-books tab.

MATERIALS AND MANUFACTURING TECHNOLOGY

METAL MATRIX COMPOSITES

J. PAULO DAVIM
EDITOR

Nova Science Publishers, Inc.
New York

For permission to use material from this book please contact us:
Telephone 631-231-7269; Fax 631-231-8175
Web Site: http://www.novapublishers.com

Additional color graphics may be available in the e-book version of this book.

Library of Congress Cataloging-in-Publication Data

Metal matrix composites / editors, J. Paulo Davim.
p. cm.
Includes bibliographical references and index.
ISBN 978-1-61209-771-8 (hardcover)
1. Metallic composites. I. Davim, J. Paulo.
TA481.M4455 2011
620.1'6--dc23
2011037648

Published by Nova Science Publishers, Inc. † New York

CONTENTS

PREFACE

Recently, the utilization of metal matrix composites (MMC's) has increased in various areas of industry due to their special mechanical and physical properties. MMC's, particularly aluminium based composites have a high specific strength and stiffness at elevated temperatures as well as fatigue, corrosion and wear resistance. Therefore MMC's have the potential to replace conventional materials in various fields of application such as automotive, aeronautical and aerospace, as well as in others advanced industries because of is own properties. This new book reviews current research on metal matrix composites.

Chapter 1 - This chapter presents the state of the art in the field of advanced metal matrix composites (MMCs) with metallic glass reinforcements. Metallic glasses are attractive as hard and strong materials more compatible with metallic matrices than ceramics due to their common metallic nature. Scarce publications appeared more than 2 decades ago presenting MMCs with amorphous reinforcements; however, it is only in the recent few years that this type of materials has gained significant interest from specialists both from the field of metastable and glassy materials and from the field of traditional MMCs. Several research groups have contributed to the development of these novel materials, so, at present, they can be truly regarded as a new class of MMCs. Proposed processing routes, microstructures obtained and mechanical properties of the prepared composites are overviewed in the chapter in order to emphasize recent developments and propose future research directions. Due to availability of a wide variety of metallic glasses, the composition, microstructure and properties of metallic glass-reinforced MMCs can be very finely tuned. Preparation of fine metallic glass particles, uniform mixing of these particles with a metal matrix and keeping the amorphous structure of the reinforcement intact upon temperature-assisted consolidation present challenges yet to be tackled in order to use the full potential of this new class of MMCs.

Chapter 2 - The MAX phases and hexagonal metals, among many other plastically anisotropic solids with c/a ratios > 1.5, have been recently classified as kinking nonlinear elastic (KNE) solids – the signature of which is the formation of fully reversible, hysteretic stress-strain loops during cyclic loadings. In this review, recent progress in a novel class of Mg matrix composites reinforced with Ti_2AlC, what is known as MAXMET, is reviewed. The Mg grains that constitute the matrix of these composites are exceptionally stable, a fact attributable to nanocrystalline nature of the Mg-matrix. This review shows that the MAXMET composites are all KNE solids and their KNE behavior is a strong function of texture, and characterized by the formation of fully reversible hysteretic stress-strain loops under uniaxial cyclic compression.

Chapter 3 - The thermal expansion of different types metal matrix composites (MMC) is investigated by dilatometry. The instantaneous linear coefficients of thermal expansion (CTE) are determined for particle, interpenetrating and short fiber reinforced (SFRM) Al-Si alloy matrices between room temperature and 500°C. Particle or short fiber preforms, infiltrated by Al-Si-casting alloys, produce 3D reinforcement architectures connected by Si-bridges. The CTE(T) for unidirectional continuous fiber reinforced (CFRM) Al- and Mg-matrices are determined in different directions between -50 and 200°C. The anisotropy in thermal expansion of SFRM and CFRM is correlated to their volumetric expansion. Predictions of thermo-elastic models are only valid as long as the matrix deforms elastically, the range of which is estimated by the volume misfit between matrix and reinforcement. Deviations are explained by plastification of the matrix and temperature dependent changes in the micro-porosity of MMC with interconnected architectures of reinforcement or unidirectional CFRM. The conservation of the volume of the constituents is compared with the experimental volume variations with temperature.

Chapter 4 - This paper presents an experimental investigation into the influence of stir casting parameters during casting of Al/5vol%Gr_p and Al/5vol%Al_2O_3-MMC. Taguchi method based design of experiment is used to optimize the stir casting process parameters for effective production of Al/Gr_p and Al/Al_2O_3-MMC,s. According to the Taguchi quality design concept an $L_{27}(3^{13})$ orthogonal array has been used to determine the S/N ratio (dB), ANOVA and 'F' test values for indicating the most significant parameters affecting the stir casting performance characteristics such as micro hardness and tensile strength of the cast metal matrix composites samples. Test results indicated that the stirring speed and stirring time are the most significant and significant casting parameters for micro hardness of Al/5vol%Al_2O_3-MMC. The interaction between pouring temperature and stirring speed is the most significant parameter for tensile strength of Al/Gr_p and Al/Al_2O_3-MMC,s. Utilizing experimental results and using Gauss elimination method mathematical models relating to the micro hardness and tensile strength are established to investigate the influence of casting parameters during stir casting of MMC,s.

Chapter 5 - In-situ titanium matrix composites have drawn great attention due to the superior properties over titanium alloys. Recent research achievements of synthesizing methods, selection of matrix and reinforcements, microstructure, mechanical properties and superplastic behavior are reviewed. The present problems are put forward and the further research fields are also discussed.

Chapter 6 - This chapter analyzes the effects of Linear Friction Welding (LFW) on the impact behaviour of an aluminium matrix composite (AA2124/25%vol.SiCp) and an unreinforced Al alloy (AA2024). The aluminum based metal matrix composite (MMC) was obtained by powder metallurgy, forged and T4 heat-treated, while the Al alloy was supplied in the extruded and T4-heat treated condition. Optical Microscopy (OM) and Scanning Electron Microscopy (SEM) with Energy Dispersive Spectroscopy (EDS) were used to characterize the effects of the welding process on the microstructural characteristics of the LFW joints, which were obtained between similar (MMC-MMC, AA2024-AA2024) and dissimilar materials (MMC/AA2024). Instrumented impact tests were carried out on Charpy specimens, machined either from the base materials or from the LFW joints. The mechanisms of failure was investigated by SEM analyses of the fracture surfaces. The microstructure of the joints was found to be dependent on both the initial microstructure of the tested materials and the welding process. The LFW induced a reduction of the impact strength of the AA2024 alloy, while it had no relevant effects on the impact strength of the AA2124/25%SiCp composite. The impact energy of the dissimilar weld was comparable to that of the MMC joint, since fracture propagated mostly on the composite side. The fracture path was usually located along the thermo-mechanically affected zone in the similar joints.

Chapter 7 - The use of metal matrix composite materials has been increased considerably over the last decade and, as a consequence, more number of researchers are working in this area. The objective of this chapter is to present a brief literature review on drilling of theses materials. Aspects such as tool materials and geometry, machining parameters and their influence on the thrust force, surface roughness, Burr formation are reviewed based on SiCp, Graphite and SiCp-Graphite reinforced Aluminium matrix composites. Additionally, the subsurface alterations and statistical analysis after drilling on these materials were discussed. This allows the readers to a better understanding of the material behaviour and drillability characteristics of this category of materials.

Chapter 8 - This chapter focuses on the multiple performances analysis in machining characteristics of drilling Hybrid metal matrix Composites produced through stir casting route. Desirability function based approach is employed for the optimization of drilling parameters namely spindle speed, feed rate and wt % of SiC based on the multiple performance characteristics including thrust force, surface roughness, and torque Experiments are conducted on Al 356- aluminum alloy reinforced with silicon carbide particles of size 25 microns and Mica of size 45 microns. Drilling test is carried out using coated HSS drill of 6 mm diameter. According to the Taguchi's quality concept, a L_{27}, 3-level orthogonal array was chosen for the experiments. An empirical model has been developed for predicting the thrust force, surface roughness and torque in drilling of Al356/SiC-Mica composites. The influences of different parameters and their interactions are studied in detail and presented in this study.

In: Metal Matrix Composites
Editor: J. Paulo Davim

ISBN: 978-1-61209-771-8

Chapter 1

METAL MATRIX COMPOSITES REINFORCED WITH METALLIC GLASS PARTICLES: STATE OF THE ART

Dina V. Dudina[1,2*], Konstantinos Georgarakis[1,3] and Alain R. Yavari[1]

[1]Science et Ingénierie des Matériaux et Procédés (SIMAP-CNRS),
Institut Polytechnique de Grenoble (INPG), rue de la Piscine,
Saint-Martin-d'Hères Campus, France
Institute of Solid State Chemistry and Mechanochemistry,
[2]Siberian Branch of Russian Academy of Sciences, Novosibirsk, Kutateladze, Russia
[3]WPI-AIMR Tohoku University, Japan Aoba-Ku, Sendai, Japan

ABSTRACT

This chapter presents the state of the art in the field of advanced metal matrix composites (MMCs) with metallic glass reinforcements. Metallic glasses are attractive as hard and strong materials more compatible with metallic matrices than ceramics due to their common metallic nature. Scarce publications appeared more than 2 decades ago presenting MMCs with amorphous reinforcements; however, it is only in the recent few years that this type of materials has gained significant interest from specialists both from the field of metastable and glassy materials and from the field of traditional MMCs. Several research groups have contributed to the development of these novel materials, so, at present, they can be truly regarded as a new class of MMCs. Proposed processing routes, microstructures obtained and mechanical properties of the prepared composites are overviewed in the chapter in order to emphasize recent developments and propose future research directions. Due to availability of a wide variety of metallic glasses, the composition, microstructure and properties of metallic

[*]Corresponding author, E-mail: dina1807@gmail.com

glass-reinforced MMCs can be very finely tuned. Preparation of fine metallic glass particles, uniform mixing of these particles with a metal matrix and keeping the amorphous structure of the reinforcement intact upon temperature-assisted consolidation present challenges yet to be tackled in order to use the full potential of this new class of MMCs.

INTRODUCTION

Traditional particle-reinforced metal matrix composites (MMCs) benefit from the combination of a metallic and a ceramic component, the latter performing a reinforcing role and increasing strength, hardness and wear resistance of the material [1-6]. However, there are always concerns of compatibility of the reinforcement and the matrix, which include wettability issues, detrimental interfacial reactions and problems of agglomeration of rigid high-melting temperature inclusions hindering densification. The difference in the thermal expansion coefficients between the metallic matrix and the ceramic reinforcement causes a build-up of residual stresses in the composite after thermal processing. Poor wettability between the phases is often a reason for the difficult-to-eliminate porosity localized at the matrix-reinforcement interfaces. In the opposite case of strong interactions, depending on their mechanical properties, interfacial reaction layers can reduce ductility. Although much effort is directed at improving the distribution of reinforcing phases in the matrix, there is often agglomeration or clustering of reinforcement particles. At high contents of reinforcing phase, it becomes even more challenging to keep the uniform distribution of inclusions in the matrix. Since ceramics are high-melting point materials, when clustered, the reinforcing particles do not sinter between themselves and, as a result, the areas of agglomeration contain a disproportionate part of the total porosity of the composite [7-8]. In this regard, it is desirable to have the matrix and the reinforcement in the composite with easy sinterability within the same temperature range in order to fully eliminate porosity. This goal can be achieved by using metallic glass reinforcements in metal matrices.

The idea of using amorphous alloy particles as a reinforcement in MMCs has attracted the attention of researchers even more than two decades ago [9-10]. In the US Patent 4, 562, 951 [9], metallic glass ribbons, wires and strips were suggested as reinforcements of extraordinary mechanical properties. A strong recommendation was given to avoid melting of the metal matrix during the processing of the composites in order not to deteriorate the properties of the metallic glass and the metal matrix itself. In order to implement a solid state processing, metallic glass strips or wires were placed between aluminum alloy discs and pressed in a hot press. The temperature of the process was high enough to induce a superplastic flow in the matrix Al alloy but still lower than the crystallization temperature of the metallic glass. Since in the described process, metallic glass was utilized in its common as-quenched shapes, such as ribbons and wires, and the question of metallic glass dispersion into particles was not addressed, this work can be regarded as the first step in forming the ideology of metallic glass-reinforced MMCs.

The same year [10], the first work on aluminum alloys reinforced with particles of metallic glass appeared. The authors used a Ni-Mo-Cr-B metallic glass powder, which they combined with a powder pre-mix of the 2014 Al alloy to further compact and sinter. The sintering conditions of the composite were not optimized: during liquid phase sintering, the compacts experienced swelling and remained porous. Though a technique of re-pressing and re-sintering of the initially sintered composites helped decrease the porosity, it was still too high: for example, in the composite with 20 vol.% of metallic glass particles, re-sintering allowed decreasing the porosity only from 39 to 33%. At that time, not much attention has been drawn to the processes occurring in the metallic glass phase itself during liquid phase sintering of the composites. The porosity of the composites made it difficult to predict or interpret their properties, such as electrical resistivity, hardness and corrosion resistance.

More than a decade later, Stawovy and Aning [11] suggested using Fe-W amorphous particles as a reinforcement for pure iron. Produced by mechanical alloying, the reinforcing particles were composed of an amorphous phase and nano-sized grains of tungsten. The sintered material was not very strong due to poor densification. No study was conducted at that time to find an optimized composition of the reinforcing amorphous phase. The ability of most amorphous metallic alloys to undergo a glass transition with a concomitant change in viscosity was not put forward as an important effect playing a decisive role for achieving pore-free composites. Seemingly, the lack of success in improving the mechanical performance of the composite has discouraged further investigations in this area for a certain period of time. Amorphous reinforcements have come back on stage in the last few years thanks to tremendous advances in the field of metallic glasses [12-16].

1. Metallic Glass Particles as an Alternative Reinforcement for MMCs

The advantage of metallic glass as a reinforcement is its softness and deformability/sinterability within the super-cooled liquid region ΔT_x and its very high strength at ambient temperatures [12]. Due to their 2% elastic strain range (as compared to 0.2% offset for crystalline materials), metallic glasses currently have the best known values for the performance index σ^2/E and good values of the index σ^2/ρ E, where σ, ρ and E are respectively the yield strength, mass density and Young's modulus. In miniature or microscopic dimensions, with high resilience and low damping coefficients, they are ideal engineering materials for wear resistant micro-gears and coatings [16]. Above their glass transition temperatures metallic glasses enter a technologically attractive Newtonian flow in the super-cooled liquid region ΔT_x between the glass transition T_g and the crystallization T_x temperatures. This temperature range $\Delta T_x = T_x - T_g$ is currently used for creating various shapes at relatively low temperatures. Like in oxide glasses, in the super-cooled liquid region ΔT_x above T_g, the viscosity of metallic glasses drops drastically and they become liquid-like.

Owing to their excellent mechanical properties at room temperature, metallic glasses are deemed to be a highly promising alternative for reinforcing metallic materials, being particularly suitable for low-melting temperature alloys, such as Al and Mg alloys. The

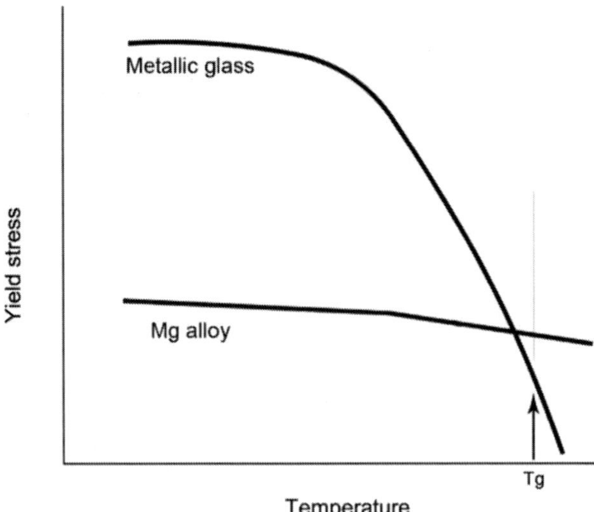

Figure 1. Schematic illustration of the change in yield stress of a metallic glass and a Mg alloy with increasing temperature, T_g is the glass transition temperature of the metallic glass [17] (Reprinted from Comp. Sci. Tech. 69, D.V.Dudina, K.Georgarakis, Y. Li, M.Aljerf, A.LeMoulec, A.R.Yavari, A.Inoue. A magnesium alloy matrix composite reinforced with metallic glass, Pages 2734-2736, Copyright (2009), with permission from Elsevier).

limitation of metallic glass as a reinforcement will lie in that it is metastable and crystallize above a certain temperature (crystallization temperature) T_x. The composite materials should be designed in such a way that the chosen sintering temperature of the composite powders is within the super-cooled liquid region ΔT_x of the glass while being very close to the solidus temperature of the matrix alloy. Above its glass transition temperature T_g, the metallic glass nano- and micron sized particles will become soft and will act as a soft binder and porosity remover. To illustrate this approach, Fig.1 schematically draws the drop in yield stress in metallic glasses and Mg alloys (as possible matrix materials) with temperature [17]. After compaction to full density and cooling to below T_g and down to room temperature, the same glassy particles become the hard reinforcement phase.

The metallic glass particles serve (i) as hard inclusions acting as obstacles for dislocation movement in the matrix and (ii) as a hard phase partially taking on the load provided there is efficient load transfer from the matrix to the reinforcement in the composite. So, the metallic glass phase is to modify the properties of the metal matrices in these composites: first of all, to increase yield strength and fracture strength. Simultaneously, a loss in ductility can be expected with the introduction of the second phase to the metals.

In a different approach, metallic glasses can play the role of a matrix and crystalline phases can constitute the second phase. In this class of composites − metallic glass matrix composites − the crystalline metal phase is usually introduced to modify the deformation behavior of single-phase metallic glasses. It is known that metallic glasses are rather brittle, bearing limited plastic strain when subjected to tensile mechanical loading at ambient temperature, due to their heterogeneous deformation and shear softening, which occurs in highly localized shear bands [15, 18-20]. In order to partially overcome this problem, other phases are introduced to induce branching of shear bands preventing their propagation and thus increasing ductility. The crystalline phase can be in the form of micron scale dendrites

[21-23] or networks [24], or in the form of nanocrystals embedded in an amorphous matrix formed either during casting [25] or partial crystallization of bulk metallic glasses BMGs [26-28].

2. Processing of Metallic Glass Particle-Reinforced MMCs

A two-phase composite structure, in which one phase is amorphous and the other is a crystalline metal or alloy, can be formed by several processing methods. In a powder state, the particles of both phases can be mixed and then sintered. Because of high sensitivity of metallic glasses to temperature, solid-state processing routes are preferable for metallic glass-reinforced composites. Consolidation of composites from the corresponding powder mixtures will be an appropriate choice to produce bulk samples.

Casting technologies for fabricating metallic glass-reinforced MMCs can be applied only in certain cases: the metallic glass should have its T_x higher than the liquidus of the matrix alloy. This requirement narrows the choice of the materials to be paired. Despite the limitations of the processing by casting, a study has been performed by Lee et al [29] showing an example, where this approach was applied successfully. A metallic glass with a high T_x and an Al alloy with good castability were chosen to produce a composite, where ribbons of metallic glass could play a role of the reinforcing phase. The metallic glass ribbons were cold pressed to make a pre-form for infiltration casting. The crystallization temperature of the metallic glass Ni–20.6Nb–40.2Ta, wt.% was higher than the liquidus temperature of the matrix alloy Al–6.5Si–0.25Mg, wt.%. Different microstructures are shown in Fig.2 [29], where the ribbons of the Ni–20.6Nb–40.2Ta are surrounded by the Al alloy matrix.

The concept of the fabrication procedure can be clearly demonstrated by the thermographs (DTA traces) of Fig. 3 [29]. The endothermic events on the curve for the composite correspond to melting of the Al alloy matrix. Crystallization of the amorphous phase is observed as an exothermic peak on the curves (marked A). The nature of the second

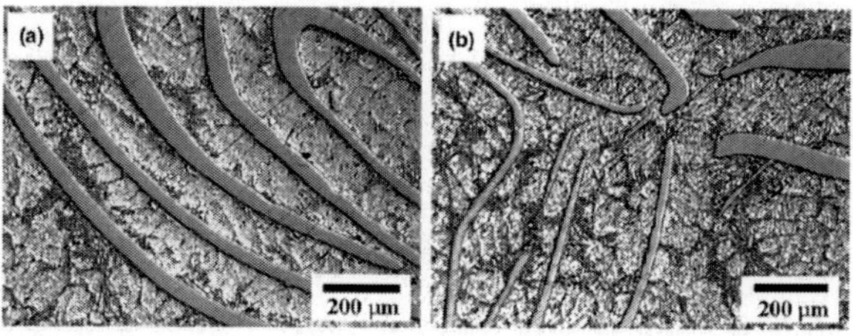

Figure 2. Optical micrographs obtained from the transverse cross-sections of the Ni–20.6Nb–40.2Ta, wt.% metallic glass ribbon-reinforced Al–6.5Si–0.25Mg alloy matrix composite; a-b – different ribbon morphologies [29]. (Reprinted from Scripta Mater. 50, M.H.Lee, J.H.Kim, J.S.Park, J.C.Kim, W.T.Kim, D.H.Kim. Fabrication of Ni–Nb–Ta metallic glass reinforced Al-based alloy matrix composites by infiltration casting process, Pages 1367-1371, Copyright (2004), with permission from Elsevier).

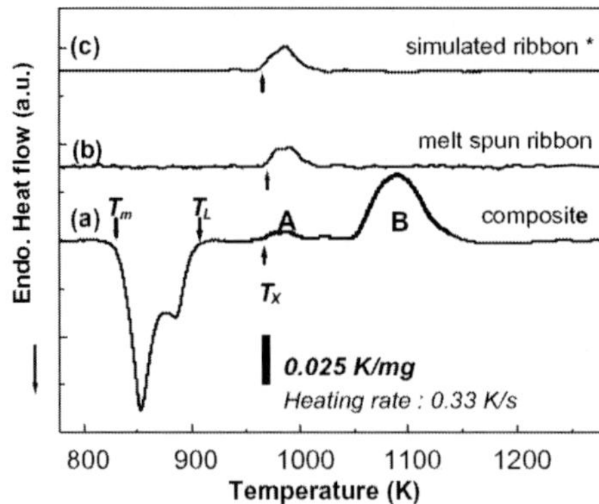

Figure 3. DTA traces obtained from the Al–6.5Si–0.25Mg, wt.% alloy matrix composites (a), melt-spun ribbon Ni–20.6Nb–40.2Ta, wt.% (b), the heat treated alloy ribbon in the same condition as the thermal cycles of the composite fabrication process, simulated ribbon (c) [29] (Reprinted from Scripta Mater. 50, M.H.Lee, J.H.Kim, J.S.Park, J.C.Kim, W.T.Kim, D.H.Kim. Fabrication of Ni–Nb–Ta metallic glass reinforced Al-based alloy matrix composites by infiltration casting process, Pages 1367-1371, Copyright (2004), with permission from Elsevier).

exothermic event (marked B) was not investigated in detail; presumably it is related to the interaction of the matrix components with those of the metallic glass at elevated temperatures. The reinforcing effect of the ribbons was reflected by an increase in the compressive yield strength from 129 MPa for the Al alloy matrix to 163MPa for the composite.

The advantage of the matrix alloy of good castability is that it can form composites without macrodefects or pores and ensure good bonding between the phases at the interface. However, if the matrix alloy melts at relatively low temperatures, this also implies that it is not very strong and may not present much interest as a suitable matrix for composites for load-bearing applications. Another difficulty of the processing through casting is lack of flexibility in varying the reinforcement content, which has to be high enough to make a porous pre-form to be infiltrated by melt. The powder metallurgy approach eliminates these problems and allows working with a variety of metallic glasses and matrix alloys utilizing solid-state sintering of the prepared powder mixtures.

A principally different approach is to form an amorphous phase in-situ in a metal matrix when an alloy of the thoroughly selected composition is solidified form the melt to form a two-phase structure. It is a known practice to use this technique to produce composites with a metallic glass matrix and dendrites of a crystalline phase seeking an increased ductility of the composite relative to the monolithic metallic glass [22-24]. Similar microstructures can be produced when the composition of the alloy is such that a high volume content of the crystalline phase makes it possible to consider it as a matrix. Hofmann et al [23] report fabrication of two-phase composites with 20 and 31vol.% of the glassy phase in the Ti-Zr-V-Cu-Al-Be system. Also, Guo et al have created a networking amorphous phase coexisting in a composite with a Ti(Zr) crystalline solid solution in the Ti-Zr-Cu-Be system [24]. The processing offers unique opportunities for making composites with titanium matrices, for which powder metallurgy is rather difficult because of relatively high temperatures required to consolidate a titanium matrix to full density; at these temperatures most metallic glass reinforcements will crystallize and lose its attractive mechanical properties transforming into brittle intermetallics.

In certain compositions, upon solidification from the melt, amorphous phases can be present in the form of nanoscale domains trapped in the grains of Al, as in the high-strength Al alloys in the Al-V-Fe system [30]. A comprehensive review of Al alloys with nanogranular amorphous phases, nanoscale networking phases and other high strength Al alloys was presented by A.Inoue [31].

An interesting solution to incorporating and dispersing metallic glass ribbons/particles in a metal matrix was proposed by Lee et al [32], who cold and warm rolled a stack of metallic glass ribbons $Ni_{59}Zr_{20}Ti_{16}Si_2Sn_3$ and sheets of Cu foil. During cycles of cold rolling, the ribbons broke into shorter pieces while their thickness remained unaffected. With an increase in the number of rolling and folding cycles, the pieces (particles) of the ribbons became smaller (Fig.4) [32]. Though beneficial for an improved distribution of metallic glass particles, repeated cold rolling caused cracks in the material due to the work-hardening effects in copper. Warm rolling that succeeded cold rolling resulted in a denser material. The temperature of the warm rolling process was chosen such that it was within the super-cooled liquid region of the metallic glass. The glassy phase did not crystallize but the particles experienced re-shaping due to a viscosity drop within the super-cooled liquid region. Fig.5, a [32] shows that the metallic glass particles have been re-shaped and no visible porosity can be seen. Fig.5, b [32] demonstrates a monolithic material obtained by warm rolling of the

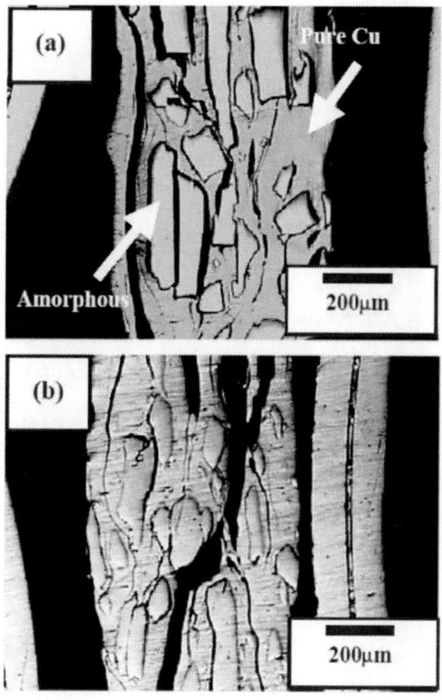

Figure 4. Optical micrographs of repeated cold rolling and folding (RCRF) cycled $Ni_{59}Zr_{20}Ti_{16}Si_2Sn_3$ ribbon/Cu composites: a - 10 RCRF processed, b - 40 RCRF processed [32] (Reprinted from Mater.Lett. 59, M.H.Lee, J.S.Park, J.H.Kim, W.T.Kim, D.H.Kim. Synthesis of bulk amorphous alloy and composites by warm rolling process, Pages 1042-1045, Copyright (2005), with permission from Elsevier).

metallic glass ribbons alone. Analogously, due to low viscosity of the metallic glass at the chosen temperature, good bonding between the ribbons helps create a monolithic material, which shows hardness comparable to that of Ni-based amorphous alloys. The successes of cold/warm rolling in dispersing metallic glass in metal matrices make it attractive for the production of dense composites simply starting with pieces of ribbons. At the same time, there are strong limitations to the possible shape and size of the final products.

The powder metallurgy approach and consolidation of metallic glass/metal composite powders allow producing bulk samples with a wider range of shapes and sizes compared to rolling processes. Indeed, consolidation of powders has been the most commonly used method to produce metallic glass-reinforced MMCs by several research groups [17, 33-45]. Warm and cold consolidation methods can be considered. As was shown by Yavari et al [46], cold consolidation via severe plastic deformation by high-pressure torsion straining can successfully transform an amorphous matrix composite with fcc Al nanocrystals into a fully dense bulk material in the Al-Fe-Nd system. It can be expected that with MMCs containing amorphous phases, this method will work equally well. Indeed, a certain success of severe plastic deformation was reported for Vitralloy106a – (50-70) vol.% W composites [36]. However, in order to beneficially use the viscosity drop in a metallic glass above its T_g to enhance sintering, thermally-assisted consolidation by hot extrusion [39-40, 42], induction heating sintering [17, 44] or Spark Plasma Sintering [45] appear to be promising for compacting metallic glass/metal composite powders.

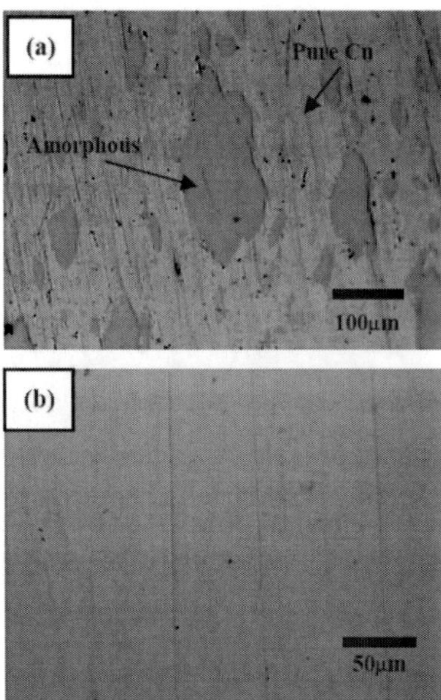

Figure 5. The warm rolling processed (a) $Ni_{59}Zr_{20}Ti_{16}Si_2Sn_3$ amorphous ribbon/Cu composite, (b) monolithic amorphous $Ni_{59}Zr_{20}Ti_{16}Si_2Sn_3$ [32] (Reprinted from Mater.Lett. 59, M.H.Lee, J.S.Park, J.H.Kim, W.T.Kim, D.H.Kim. Synthesis of bulk amorphous alloy and composites by warm rolling process, Pages 1042-1045, Copyright (2005), with permission from Elsevier).

3. METALLIC GLASS PARTICLE-REINFORCED MMCS WITH DIFFERENT METAL MATRICES

This section is aimed to review the experimental results obtained in the past few years on the processing and properties of the glassy alloy particle-reinforced composites with different matrices. Research effort was mainly directed to aluminum matrix composites as aluminum is suitable for numerous applications. Due to availability of Al alloys with known mechanical properties, they were often selected as matrices rather than pure aluminum. An attempt was also made to develop a Mg alloy matrix composite reinforced with metallic glass dispersoids. Other metals with melting temperatures higher than those of Mg and Al alloys have also been evaluated (Cu, Fe, Ti, Ni, W) as potential matrices for amorphous phase-reinforced MMCs. In these cases, one has to resort to special procedures to produce a bulk composite.

3.1. Aluminum and Its Alloys Reinforced with Metallic Glass

Al and its alloys have been the most widely used matrix materials in MMCs, both in research and industrial applications due to low density, lower cost compared to magnesium and titanium alloys, and significant strength, ductility and corrosion resistance [1-6]. The strength of aluminum is nonetheless to be improved and much effort has been directed to the development of aluminum matrix composites or high-strength aluminum alloys [1-3, 5, 31]. Aluminum can accommodate many reinforcements, such as silicon carbide, boron carbide, aluminum oxide and titanium diboride. The common problems encountered in ceramic particle-reinforced aluminum are interfacial decohesion, remaining porosity and undesirable

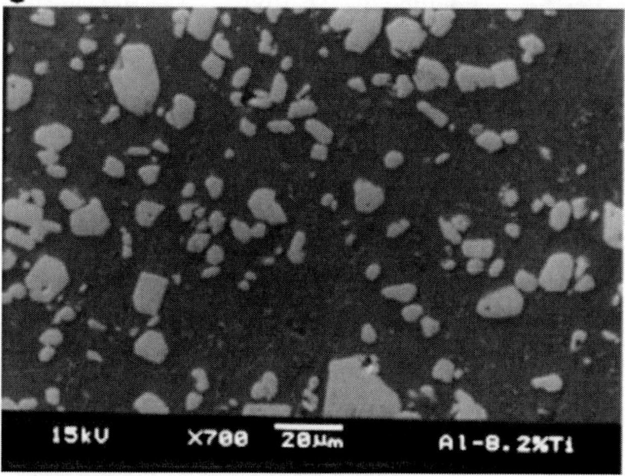

Figure 6. SEM micrograph showing the uniform distribution of Ti particle reinforcement in the Al–7.5%Ti composites [48] (Reprinted from Composites Part A 38, S.K.Thakur, M.Gupta. Improving mechanical performance of Al by using Ti as reinforcement, Pages 1010-1018, Copyright (2007), with permission from Elsevier).

reactions at the interfaces. Ceramic particles have to be first coated with a metallic layer before their introduction into the matrix to improve densification and mechanical properties of the composites [47].

In order to overcome difficulties related to the difference in the mechanical and physical properties of a metal matrix and a ceramic reinforcement, titanium as a stronger metal was suggested for the role of reinforcement aiming at increased mechanical strength without dramatic ductility losses [48]. Fig.6 shows the microstructure of titanium-reinforced aluminum with titanium particles distributed in the Al matrix produced by disintegrated melt deposition.

Another possibility of improving the mechanical behavior of aluminum is to synthesize alloys with special microstructures. A wide variety of nano-crystalline and nanoquasicrystalline Al-based alloys has been developed by the rapid solidification technique through controlling the composition of the super-cooled liquid and the processing parameters, as reviewed by A.Inoue [31]. Aluminum alloys with quasicrystalline phases show good ductility and increased mechanical strength possessing a unique structure consisting of an icosahedral phase distributed in the fcc Al. In the Al-Mn-Ce-Co alloy, the value of tensile strength can reach 600-800 MPa while ductility varies in the range of 5-10%. The Al-Cr-Ce-Co extruded samples produced from icosahedral base atomized powders show lower strength of about 550 MPa remaining more ductile (up to 24%) [49]. Alloys with good mechanical strength at elevated temperatures can be produced in Al-Fe-Cr-Ti system [50]. The improvement of thermal stability still remains a challenge in the development of such alloys. The promising approach is based on varying the composition of the icosahedral phase [51-52].

Al-based chill-zone alloys with very high strength (>1 GPa) comparable to that of fully amorphous Al alloys and good plasticity in compression have been recently produced by Li et al in the Al-Ni-Y system by graphite mold casting [53-54]. The alloys did not contain amorphous phases but consisted of the fcc Al solid solution and $Al_{19}Ni_5Y_3$ metastable phase. This combination of

properties was achieved due to a bimodal-type structure of the alloys having a submicrocrystalline surface layer and an inner core with a coarser micro-stucture .

A possibility of using intermetallic compounds as reinforcements in Al alloys was explored by Zhang et al [55]. An unusual method of introducing particles of intermetallics was suggested based on in-situ devitrification of an amorphous alloy powder mixed together with the matrix Al alloy powder. The composition of the amorphous alloy – $Al_{85}Ni_{10}La_5$ – was chosen such that it could give thermodynamically stable phases upon complete devitrification (Al, $Al_{11}La_3$, Al_3Ni). The main idea of this work was to keep the fine size of the dispersions during the conventional processing by cold isostatic pressing and extrusion. The choice of the reinforcement composition was made to provide better compatibility between the matrix and reinforcement both having metallic chemical bonds. The amorphous alloy in this work was not intended to be used in its glassy state; rather, it was deliberately crystallized to yield fine intermetallic reinforcements. This concept has its advantages in the sense of making composites of certain thermal stability within the chosen temperature range.

In spite of a great number of studies already performed on improving the properties of aluminum matrix composites, there is still constant interest to novel reinforcement options and new processing methods possibly resulting in better mechanical performance of the composites. For each alloy or composite, several factors have to be taken into account such as microstructure, resultant mechanical properties and raw material and processing costs, so there is always room for novel compositions or optimization of manufacturing methods.

The advances in the field of metallic glasses opened up opportunities for making amorphous alloys with extremely high strength [13-15]. However, metallic glasses suffer from localized shear [15, 18-20] and limited ductility under mechanical load – one of the main reasons their potential as structural materials has not yet been fully used. Led by the idea to utilize the excellent mechanical properties of metallic glass in structural applications, several research teams embarked on the studies of composites containing metallic glass particles as inclusions in a softer phase. The ductility lacking in metallic glasses is to be given by a metal matrix, in which metallic glass is evenly distributed.

In the powder metallurgy processing of metallic glass-reinforced MMCs, there are two possibilities of obtaining a dispersion of metallic glass particles in the metal matrix: the powders of the matrix metal or alloy can be mixed with the selected amorphous powders or the amorphous phase can be produced by a solid-state in-situ reaction in a mixture of elemental powders with an excess of the matrix metal. The latter approach has been used by Botta et al [56] for the Al-Fe-Zr system, in which an amorphous phase formed during mechanical alloying while the residual Al remained crystalline. Further, this composite powder mixture was consolidated by severe plastic deformation under high-pressure torsion straining and a certain amount of the amorphous phase remained in the compacted material.

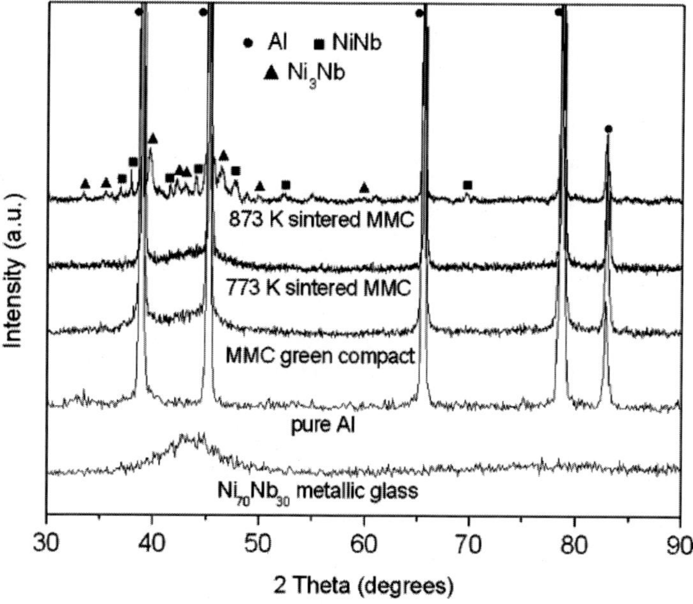

Figure 7. XRD patterns for pure Al, $Ni_{70}Nb_{30}$ glassy powders, the Al–30 wt.% $Ni_{70}Nb_{30}$ green compact, and the Al–30 wt.% $Ni_{70}Nb_{30}$ MMC sintered at 773 K and 873 K for 2 h [35] (Reprinted from Scripta Mater. 54, P.Yu, K.B.Kim, J.Das, F.Baier, W.Xu, J.Eckert. Fabrication and mechanical properties of Ni-Nb metallic glass particle-reinforced Al-based metal matrix composite, Pages 1445-1450, Copyright (2006), with permission from Elsevier).

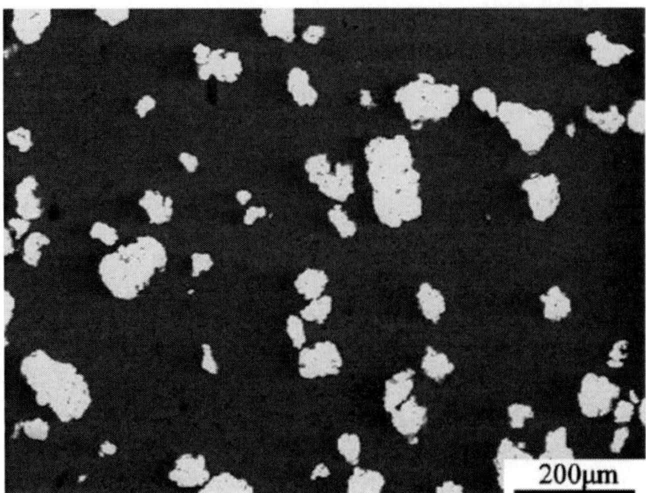

Figure 8. SEM micrograph for the Al–30 wt.% $Ni_{70}Nb_{30}$ sample sintered at 773 K for 2 h. [35] (Reprinted from Scripta Mater. 54, P.Yu, K.B.Kim, J.Das, F.Baier, W.Xu, J.Eckert. Fabrication and mechanical properties of Ni-Nb metallic glass particle-reinforced Al-based metal matrix composite, Pages 1445-1450, Copyright (2006), with permission from Elsevier).

Mixing in the solid state of two phases prepared separately appears to be easier in terms of controlling the composition of the amorphous phase in the composite. Understandably, researchers have started with a simple binary amorphous alloy to ex-situ introduce it in a metal matrix and evaluate its behavior in the composite [35]. The $Ni_{70}Nb_{30}$ alloy was chosen as a metallic glass with a relatively high crystallization temperature (T_x=816 K) and high strength. The alloy was prepared in the powder form by mechanical alloying and did not require any further dispersion.

The sintering of the mixture of the Al powder and the amorphous $Ni_{70}Nb_{30}$ powder was performed at a temperature (773 K) lower than the crystallization temperature of the amorphous phase. The XRD (Fig.7, [35]) and DSC analyses confirmed the presence of the glassy reinforcement in the composite. The particles of the glassy reinforcement were uniformly distributed in the microstructure of the sintered composite (Fig.8, [35]). The ultimate compressive strength of the composite is found to be higher than that of the pure bulk aluminum as can be seen in Table 1 in the end of this section.

The thermal stability of the metallic glass phase does not only concern its crystallization; interactions between the phases may take place at the interfaces. Mechanical properties of the composites depend greatly on the processes at the interfaces, which can influence the load transfer from the matrix to the reinforcement. Interactions at the interface occurring to some extent may improve the bonding while interactions leading to the formation of product layers can degrade the properties of the interface and the composite as a whole. In the case of multicomponent metallic glasses, the situation may be quite complicated because reactivity and diffusivity of several elements should be taken into account. As a result of the complex chemistry of the reinforcement, the products of interaction can contain multiple phases. So far, very few investigations have been carried out on the interfacial reactions in metallic glass-reinforced metals. Thanks to the distinct works of Yu et al [38-39], it becomes clear that significant interaction between the amorphous reinforcement and the metal matrix resulting in the formation of intermetallic products will most likely take place after the crystallization of the metallic glass is completed, though this can happen well below the melting temperature of the matrix. Yu et al [38] have shown that during isothermal annealing, the crystallization of the $Ni_{60}Nb_{40}$ amorphous reinforcement precedes its interaction with the Al matrix. The interaction is diffusion-controlled and results in the formation of two intermetallic layers (Fig.9, 10). Lee et al [29] made a similar observation of a reaction between amorphous ribbons used as reinforcements and an Al alloy matrix, which occurred only after the crystallization of the metallic glass had been completed. Consequently, when the composites are processed by powder metallurgy and sintered, the formation of the intermetallic layers that are likely to form at the metallic glass/matrix interfaces and/or interdiffusion of the metal components between the phases should be taken into account.

The Al-30 wt.%$Ni_{60}Nb_{40}$ composites show better mechanical performance after hot pressing and hot extrusion as compared to the composites conventionally sintered from cold-pressed green bodies at the same temperatures [39]. The reason for this is a density increase in the composites consolidated using pressure-assisted methods. An interesting observation was made regarding the thermal behavior of the composites consolidated using hot pressing, hot extrusion and conventional sintering. During annealing of the consolidated samples at a temperature allowing crystallization of the amorphous phase and its reaction with the matrix, the interaction of the Al matrix and the reinforcement starts earlier in time in denser samples, i.e. in the samples produced by hot pressing and hot extrusion. The observed effect was rationalized by an easier mass transfer across the matrix-reinforcement interface [39]. Also, the authors assumed that the influence of high pressures during consolidation was reflected in a stronger tendency of the metallic glass in the composite to crystallize, which, in turn speeds up the interaction of the reinforcement and the matrix, as this interaction occurs much faster after the amorphous reinforcement has been crystallized.

Consolidation using only high pressures, without a suitable heat treatment, appears to be insufficient to produce a good composite. Explosive compaction of an Fe-based metallic glass powders mixed with Al and Al-Ni powders allows producing a bulk composite keeping the metallic glass amorphous [43]. The problems of the composite were the remaining porosity and brittleness. The authors argue that high pressures in the explosive compaction create high-temperature regions within the sample. As a result, the particle surfaces experience melting followed by rapid solidification that can help preserving the amorphous state of the reinforcement. However, the volume fraction of the material that can experience melting is very small compared to the total volume of the compact, so the resultant material lacks good bonding between the powder particles and remains porous.

A challenging task was undertaken by Scudino et al [40] and Samanta et al [37], who made attempts to reinforce pure Al with Al-based metallic glasses, which are known for their rather low crystallization temperatures. Obviously, the consolidation of the composite from the powder state had to be performed at low temperatures as well. Samanta et al [37] prepared a mixture of the pure Al powder and the $Al_{80}Ni_{10}Ti_{10}$ amorphous powder, the latter produced by mechanical alloying. In order to obtain a bulk composite, compaction at a high pressure of 7 GPa within an Al-container with subsequent rolling was applied. Despite the constraint for the allowed temperature, Scudino et al [40] produced bulk composites containing the $Al_{85}Y_8Ni_5Co_2$ metallic glass as a reinforcement with relative densities higher

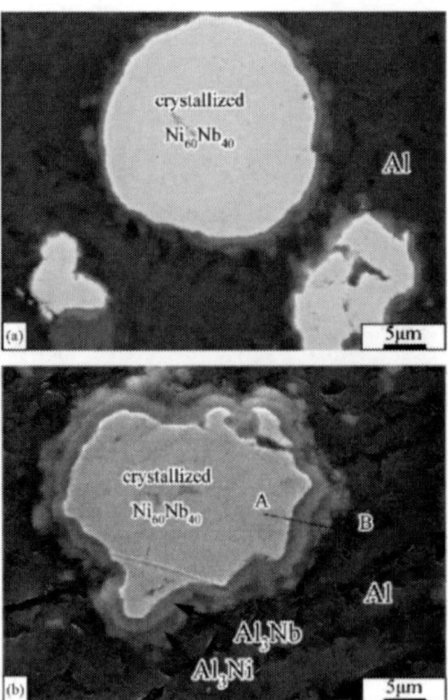

Figure 9. SEM micrographs of Al–30 wt.% $Ni_{60}Nb_{40}$ MMCs annealed at 913K for (a) 800 s and (b) 900 s [38] (Reprinted from Mater. Sci. Eng. A 444, P.Yu, L.C.Zhang, W.Y.Zhang, J.Das, K.B.Kim, J.Eckert. Interfacial reaction during the fabrication of $Ni_{60}Nb_{40}$ metallic glass particles-reinforced Al based MMCs, Pages 206-213, Copyright (2007), with permission from Elsevier).

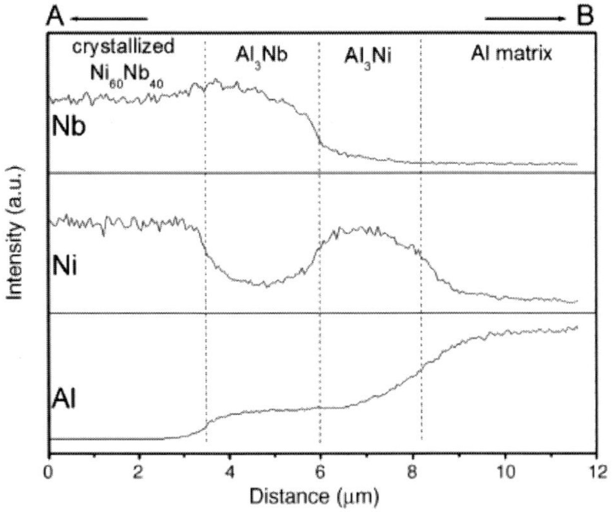

Figure 10. EDX line scan for the Al–30 wt.% $Ni_{60}Nb_{40}$ MMC annealed at 913K for 900 s showing plots of the intensities of Al, Ni and Nb along the length of the line AB labeled in Fig.9, b [38] (Reprinted from Mater. Sci. Eng. A 444, P.Yu, L.C.Zhang, W.Y.Zhang, J.Das, K.B.Kim, J.Eckert. Interfacial reaction during the fabrication of Ni60Nb40 metallic glass particles-reinforced Al based MMCs, Pages 206-213, Copyright (2007), with permission from Elsevier).

than 97% for the Al+50 vol.%$Al_{85}Y_8Ni_5Co_2$ composite and higher than 99% for the Al+30vol.%$Al_{85}Y_8Ni_5Co_2$ composite. The authors carefully measured the viscosity of the metallic glass with increasing temperature and found that it drops drastically at temperatures above 520 K, which was chosen as the temperature for extrusion. The extruded composites were much stronger than the pure Al matrix processed under the same conditions (Table 1).

The composites showed a behavior best described by the iso-stress model (Reuss model), which presents the lower bound for the properties of a mixture of two phases [57]. The authors believe that a peculiar orientation of the metallic glass particles, which are flaky in morphology, can be responsible for the observed behavior. The flakes lie in parallel planes within the composite and during the compression tests this plane is normal to the loading direction. This makes the material's behavior similar to that of short fiber reinforced composites, when they are loaded perpendicular to the direction of fibers.

The mechanical behavior of Zr-based metallic glass particle-reinforced aluminum matrix composites was studied by Scudino et al [42]. In this work, the analysis of the Al-$Zr_{57}Ti_8Nb_{2.5}Cu_{13.9}Ni_{11.1}Al_{7.5}$ composite with the metallic glass reinforcement content 40 and 60 vol.% was performed. These composites did not follow the rule of mixtures, their yield strength being much lower than predicted. The reason for this was the interaction between the particles and the effect of particle contiguity. This work clearly demonstrates that metallic glass/metal composites with high volume fraction of the amorphous phase require a more sophisticated analysis to predict their mechanical behavior.

Substantial reinforcing effects were achieved when the alloy Al520 was strengthened by a Cu-based metallic glass [44]. The Al alloy was produced by casting and further melt-spun to obtain ribbons that could be easily cut in pieces. The $Cu_{54}Zr_{36}Ti_{10}$ metallic glass ribbons were produced by melt-spinning and then were also cut in pieces and milled. Milling produced the $Cu_{54}Zr_{36}Ti_{10}$ powder with the amorphous structure. This powder was mixed with pieces of the alloy and milled. The composite was sintered by induction heating in a steel die under a pressure of 50 MPa at the maximum temperature of 720 K (which was within the super-cooled liquid region of the metallic glass) with the holding time at this temperature 2 min. Note that the pressure during sintering was low and the holding time was very short. Nevertheless, as can be seen from the SEM results, the composite is fully dense (Fig.11, [44]). The material features a uniform distribution of the reinforcement particle in the matrix. The bright particles are those of the $Cu_{54}Zr_{36}Ti_{10}$ metallic glass while the dark phase corresponds to the Al alloy as confirmed by the EDS analysis (Fig.11, a).

The sintered composite showed yield strength about 3 times higher than that of the matrix alloy (Fig.12 [44] and Table1) and about 35% higher than that predicted by the upper bound given by the rule of mixtures (iso-strain model) [57]. In this case, the particles of the glassy reinforcements are well separated from each other by the matrix and do not interact between themselves, rather, they interact with the matrix. The significant increase in mechanical strength of the composites can be attributed to the load transfer from the matrix to the glassy particles, which results in shifting mechanical yielding of the matrix to higher stresses, and to the strengthening effect of particles, which behave as obstacles to the movement of dislocations within the crystalline matrix.

Table 1 compares the compressive mechanical properties of the Al alloy matrix composites reinforced with metallic glass particles with those of the composites reinforced with silicon carbide particles [58-59] and carbon fibers [60]. Evidently, comparable or higher levels of mechanical strength gain can be achieved when metallic glass plays a role of reinforcement. The Al520+15 vol.%$Cu_{54}Zr_{36}Ti_{10}$ composite is remarkable in the sense that it was produced by powder metallurgy and sintering but competes well with the cast Al alloy composites strengthened by SiC particles [59].

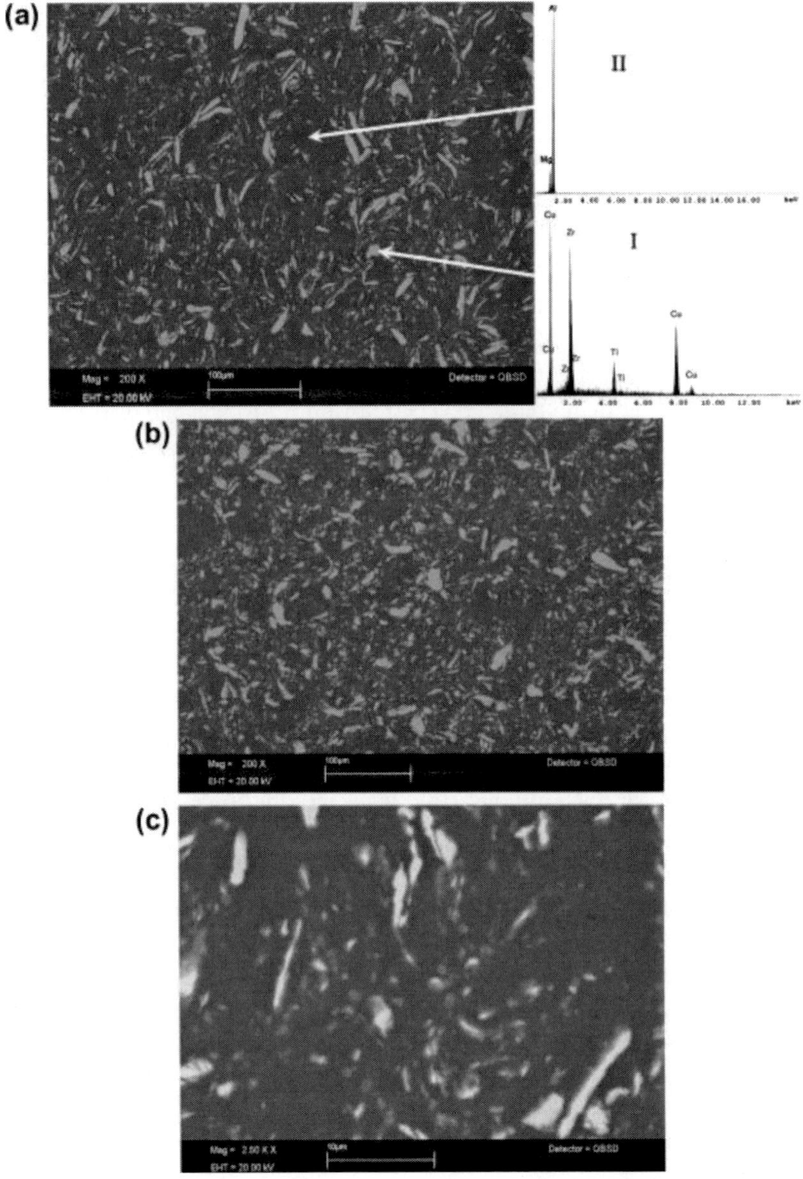

Figure 11. Microstructure of the sintered Al alloy–15 vol.% glassy $Cu_{54}Zr_{36}Ti_{10}$ composite (cross-sections of the sintered specimen, back-scattered electron images): (a) the image plane is parallel to the pressing direction; insets: representative EDS spectra form the bright (I) and the dark (II) colored areas corresponding to the Cu based glassy particles and the Al matrix alloy respectively, (b) the image plane is normal to the pressing direction, (c) image plane as in (b) – higher magnification image [44] (Reprinted from Composites Part A 41, D. V. Dudina, K. Georgarakis, Y. Li, M. Aljerf, M. Braccini, A. R. Yavari, A. Inoue. Cu-based metallic glass particle additions to significantly improve overall compressive properties of an Al alloy, Pages 1551-1557, Copyright (2010), with permission from Elsevier).

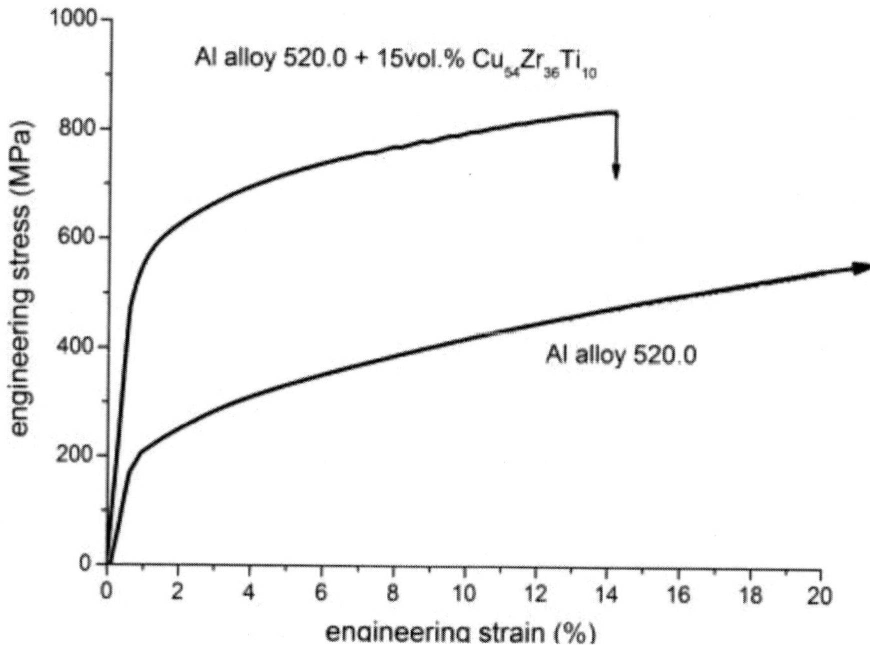

Figure 12. Compression stress–strain curves for the sintered Al alloy–15 vol.% glassy $Cu_{54}Zr_{36}Ti_{10}$ composite and the cast Al alloy alone [44] (Reprinted from Composites Part A 41, D. V. Dudina, K. Georgarakis, Y. Li, M. Aljerf, M. Braccini, A. R. Yavari, A. Inoue. Cu-based metallic glass particle additions to significantly improve overall compressive properties of an Al alloy, Pages 1551-1557, Copyright (2010), with permission from Elsevier).

The Al520+15 vol.%$Cu_{54}Zr_{36}Ti_{10}$ composite offers significant weight reduction of a part made of it compared to the situation when the part is made of the Al520 alloy. If a material is stronger, a smaller cross-sectional area of the part can be used in the design keeping the length of the part constant. A stronger composite material, as a rule, has a higher mass density than the material being substituted for, which, as we assume in our case, is the matrix alloy. If we replace Material A by Material B in a structure, then the relative weight reduction can be calculated as follows: $(w_A - w_B)/ w_A = 1- w_B/ w_A$, where w_A and w_B are the masses of Materials A and B, respectively. As was calculated by Dudina et al [44] using the yield strength values of the Al520 alloy and the Al520+15 vol.%$Cu_{54}Zr_{36}Ti_{10}$ composite, the relative weight reduction in this case is 58.3%. The high percent of the relative weight reduction can be rationalized by taking into account significant gains in yield strength with the introduction of the metallic glass reinforcing particles along with only a slight increase in the mass density. Since the properties of the composite are inevitably dictated by the processing used for its fabrication, to possess the values of strength used in these calculations, this composite (and the part in question) has to be made by powder metallurgy and presumably substitute for the matrix cast alloy. According to size specifications for components produced by powder metallurgy, they range from 1 mm to 300 mm, so their total weight lies in the range from 1 g to 2 kg [61]. So, the relative weight reduction should be considered in connection with this weight range.

Table 1. Mechanical properties in compression of Al and its alloys reinforced with metallic glass particles in comparison with Al matrix composites with other types of reinforcements

Composite or matrix material	Processing	Strain rate, s^{-1}	Yield strength, MPa	Ultimate strength, MPa	Strain at fracture %	Ref.
Pure Al	hot pressing + hot extrusion	8×10^{-4}	-	155	-	[40]
Al356+20 vol.%(Ni-Nb-Ta) amorphous ribbons	infiltration casting	4×10^{-3}	163	320	16	[29]
Al+30 wt.%$Ni_{70}Nb_{30}$	sintering	8×10^{-4}	111	146	-	[35]
Al+30 wt.%$Ni_{60}Nb_{40}$	hot extrusion	8×10^{-4}	134	-	-	[39]
Al+30 vol.%$Al_{85}Y_8Ni_5Co_2$	hot extrusion	8×10^{-4}	120	255	-	[40]
Al+50 vol.%$Al_{85}Y_8Ni_5Co_2$	hot extrusion	8×10^{-4}	130	295	-	[40]
Al+40 vol.%$Zr_{57}Ti_8Nb_{2.5}Cu_{13.9}Ni_{11.1}Al_{7.5}$	hot pressing + hot extrusion	8×10^{-4}	-	200	70	[42]
Al+60 vol.%$Zr_{57}Ti_8Nb_{2.5}Cu_{13.9}Ni_{11.1}Al_{7.5}$	hot pressing + hot extrusion	8×10^{-4}	-	250	40	[42]
Al5083+10 vol.% $Al_{85}Ni_{10}La_5$ (fully crystallized)	hot extrusion and swaging	10^{-3}	729	-	22.5	[55]
Al520+15 vol.%$Cu_{54}Zr_{36}Ti_{10}$	induction heating sintering under pressure	10^{-3}	580	840	14	[44]
Al520	casting	10^{-3}	190	-	-	[44]
Al2024+40 vol.% SiC_p	squeeze casting	10^{-3}	-	450	6.6	[58]
Al8090+12 vol.% SiC_p	stir casting	10^{-3}	381	740	25.1	[59]
(Al-Mg) alloy+30vol.% PAN-based carbon fibers	casting	10^{-3}	350-450	520-550	18-20	[60]
Al5083+5.3 vol.%B_4C	cold isostatic pressing+ extrusion	-	354	534	55	[47]

3.2. Magnesium Alloys Reinforced with Metallic Glass

The properties of magnesium - low mass density, high specific strength, good castability and weldability – make it an attractive matrix for metal matrix composites [2, 5, 62]. Compared to polymeric materials, magnesium has better mechanical properties, better electrical and thermal conductivity and it is recyclable. The inherent drawbacks of magnesium include high chemical reactivity, low elastic modulus, limited strength and creep resistance at elevated temperatures as well as low wear resistance. However, magnesium can be reinforced with ceramic particles, such as MgO, SiC, TiC, Mg_2Si, which greatly improve its mechanical behavior [2, 5, 62].

A series of studies devoted to alternative reinforcements for magnesium – stronger metals – have been recently performed in a pursuit of increasing strength while maintaining ductility [63-67]. Indeed, particles of stronger metals can impart strength to the composite keeping appreciable levels of ductility. Of course, one cannot anticipate a significant mechanical strength gain from reinforcements that are not too strong themselves and cannot win the game over ceramic reinforcements.

Until now, there has been only a single report on magnesium matrix composites reinforced with metallic glass particles. A Mg alloy- $Zr_{57}Nb_5Cu_{15.4}Ni_{12.6}Al_{10}$ metallic glass composite has been prepared by Dudina et al [17]. The matrix for the composite was the commercial Mg alloy AZ91 and the reinforcement was the well-known commercial metallic glass Vitralloy 106 $Zr_{57}Nb_5Cu_{15.4}Ni_{12.6}Al_{10}$. The Vitralloy 106 was produced in the ribbon shape and the matrix alloy was also melt-spun to produce ribbons. In the latter case, the cooling rate was not a critical parameter since melt-spinning was performed only to prepare a shape that is convenient to perform subsequent dispersion of the alloy down to powder by milling. Pieces of ribbons of the metallic glass were first separately milled to produce an amorphous powder, which was further mixed with the pieces of ribbons of the matrix alloy and the mixture was milled to produce a composite powder. After consolidation by induction heating sintering in a steel die at a temperature within the super-cooled liquid region of the metallic glass under a pressure of 50 MPa, the glassy reinforcement was still amorphous as was confirmed by the halo in the XRD pattern around $2\theta \approx 38$ degrees (Fig.13, [17]). The other two phases in the composite are Mg and the $Mg_{17}Al_{12}$ intermetallic compound as expected for the AZ91 Mg alloy. A micrograph on Fig.14 [17] shows the microstructure of the sintered composite.

The first two lines in Table 2 compare mechanical properties of the sintered Mg alloy-15 vol.% $Zr_{57}Nb_5Cu_{15.4}Ni_{12.6}Al_{10}$ composite and the matrix Mg alloy produced by casting. The results are fascinating in the sense that the sintered composite is much stronger than the matrix alloy produced by casting. Worth noting is the appreciable ductility of the sintered composite. Also presented in Table 2 are the corresponding mechanical properties in compression of Mg and Mg alloy matrix composites reinforced with ceramic particles. The ultimate strength and fracture strain of the sintered Mg alloy-15 vol.% $Zr_{57}Nb_5Cu_{15.4}Ni_{12.6}Al_{10}$ are comparable with the ceramic-reinforced composites with Mg alloy matrices while its yield strength is much higher.

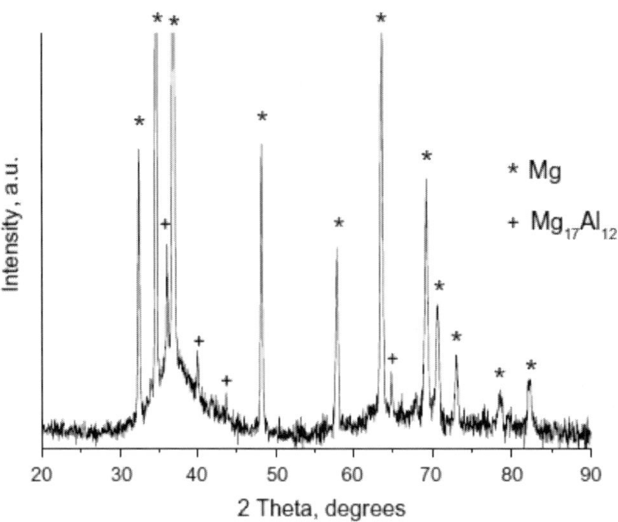

Figure 13. XRD pattern of the sintered Mg alloy–15 vol.% $Zr_{57}Nb_5Cu_{15.4}Ni_{12.6}Al_{10}$ metallic glass composite [17] (Reprinted from Comp. Sci. Tech. 69, D.V.Dudina, K.Georgarakis, Y. Li, M.Aljerf, A.LeMoulec, A.R.Yavari, A.Inoue. A magnesium alloy matrix composite reinforced with metallic glass, Pages 2734-2736, Copyright (2009), with permission from Elsevier).

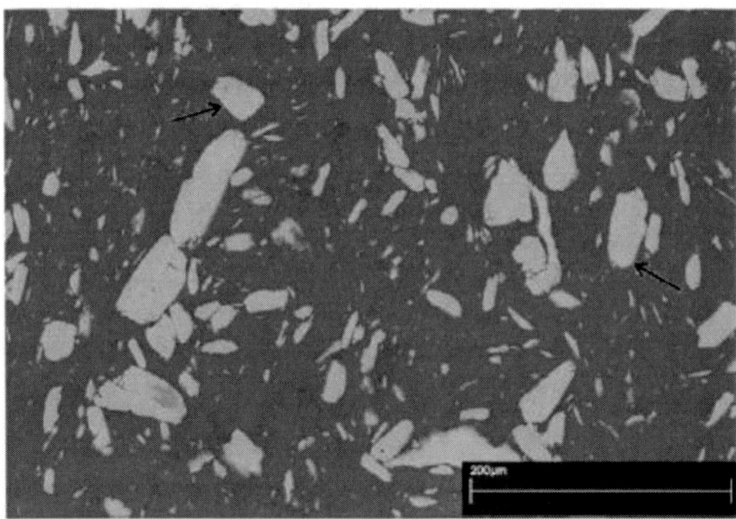

Figure 14. Microstructure of the sintered Mg alloy– 15 vol.% $Zr_{57}Nb_5Cu_{15.4}Ni_{12.6}Al_{10}$ metallic glass composite (back-scattered SEM image, the cross-section of the sintered specimen, the metallic glass particles are marked with arrows) [17] (Reprinted from Comp. Sci. Tech. 69, D.V.Dudina, K.Georgarakis, Y. Li, M.Aljerf, A.LeMoulec, A.R.Yavari, A.Inoue. A magnesium alloy matrix composite reinforced with metallic glass, Pages 2734-2736, Copyright (2009), with permission from Elsevier).

Table 2. Mechanical properties of some Mg and Mg alloy matrix composites in compression

Composite	Processing	Hardness, HV	Strain rate, s^{-1}	Yield strength, MPa	Ultimate strength, MPa	Strain at fracture, %	Ref.
Mg AZ91– 15 vol.% $Zr_{57}Nb_5Cu_{15.4}Ni_{12.6}Al_{10}$ metallic glass	induction heating sintering	123	10^{-3}	325	542	10.5	[17]
Mg AZ91	casting	68	10^{-3}	143	404	21.0	[17]
Mg AZ91 -10 wt.%SiC – 3 wt.%Si	squeeze casting	-	3.3×10^{-5}	150	350	12	[68]
Mg - 15.wt.% hydroxyapatite	melting + extrusion	-	2×10^{-4}	147.1	298.2	18	[69]
Mg ZM61 - 15.wt.% hydroxyapatite	melting + extrusion	-	2×10^{-4}	245.3	388.3	13	[69]
Mg AZ31B -3.3 wt.%Al_2O_3-10wt.%Cu	disintegrated melt deposition technique	-	-	235	530	10.7	[70]
Mg AZ31B -3.3 wt.%Al_2O_3-18wt.%Cu	disintegrated melt deposition technique	-	-	260	550	9.8	[70]

3.3. Other Metals Reinforced with Metallic Glass

It is indeed very tempting to combine metallic glass reinforcements with other metals and alloys as matrices. This section of the chapter will review several examples of metallic glass-reinforced composites with Ti, Cu, Ni, Fe and W matrices.

Until now, no composite has been reported with a titanium matrix and a metallic glass reinforcement fabricated by a powder metallurgy route. However, as was already mentioned in the previous sections of this chapter, the route to such composites lies in the processing based on solidification of alloys under a proper temperature regime [23-24].

Another popular and attractive metal matrix material is copper, understandably owing to its high electrical conductivity and good ductility. Because of a much higher melting temperature of Cu than that of Al and Mg, it becomes challenging to produce a good composite material being limited in temperatures allowed for sintering or other thermally-assisted consolidation methods. The problem of remaining porosity seems to be inevitable if a metallic glass powder mixed with a copper powder is sintered. There may be a partial solution, which uses high ductility of copper and its ability to easily deform under a proper mechanical processing. A distinct example is the work of Lee et al [32] discussed earlier in this chapter, which showed that a dense composite material is possible to produce using warm rolling of a mixture of the crystalline copper foil and pieces of the $Ni_{59}Zr_{20}Ti_{16}Si_2Sn_3$ metallic glass.

Though thought of as most suitable for low melting temperature metals, amorphous reinforcements have been combined also with an iron matrix as was mentioned earlier in this chapter. An attempt was made to reinforce a pure iron matrix with particles of an amorphous Fe-W phase as a hard reinforcement chemically compatible with the matrix [11]. The reinforcing phase particles were produced by mechanical alloying; they were not fully amorphous but consisted of an amorphous Fe-W phase mixed with nanocrystalline tungsten (Fig.15, [11]). Unfortunately, the composites were very porous (Fig.16, [11]), the porosity increasing with the volume content of the Fe-W reinforcement phase.

A similar processing route was used much more recently by Wensley et al [34] for a Ni matrix reinforced with a Ni-W amorphous phase. A two-stage mechanical milling was used to prepare the composite powder. At the first stage, the amorphous Ni-W reinforcement was produced, and at the second stage, it was milled with an additional amount of crystalline Ni to produce the desired compositions with 5-25 vol.% of the amorphous phase. Bulk samples were made by hot-isostatic pressing.

A group of metallic glass/metal composites containing 50-70 vol.% of W has drawn attention recently [36, 45, 71-72]. This might seem as an unexpected turn, since in the above sections the preferable use of low melting metals as matrices with amorphous reinforcements was elaborately justified. However, the story of W-metallic glass composites is rather interesting: they have been approached with two different goals. One of them comes from the fundamental interest to deformation mechanisms in metallic glass matrix composites. Tungsten can be added to metallic glass to create interfaces making branching of shear bands possible [36, 71-72]. The second goal explaining the interest to these composites is to create materials for kinetic energy penetrators based on tungsten as a heavy metal. In this case, the

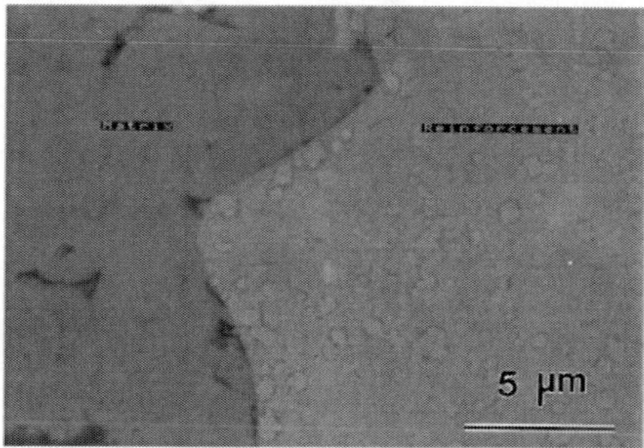

Figure 15. SEM micrograph of the two-phase reinforcement surrounded by the crystalline Fe matrix. The reinforcement particle shows a two-phase structure with tungsten rich sub-particles surrounded by an iron rich phase [11] (Reprinted from Mater. Sci.Eng. A 256, M.T.Stawovy, A.O.Aning. Processing of amorphous Fe-W reinforced Fe matrix composites, Pages 138-143, Copyright (1998), with permission from Elsevier).

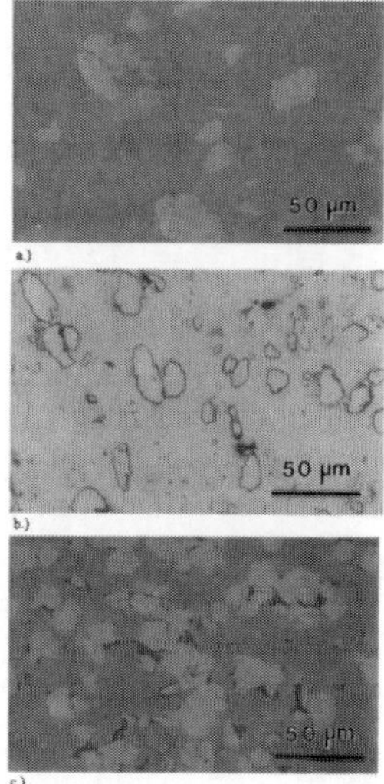

Figure 16. Micrographs of Fe matrix reinforced with Fe(W)-W particles: a - SEM micrograph of the 8.2 vol.% reinforcement sample processed using an attritor mill, b - Optical micrograph of the 8.2 vol.% reinforcement sample processed using a SPEX™ mill, c - SEM micrograph of the 33.7 vol.% reinforcement sample processed using an attritor mill [11] (Reprinted from Mater. Sci.Eng. A 256, M.T.Stawovy, A.O.Aning. Processing of amorphous Fe-W reinforced Fe matrix composites, Pages 138-143, Copyright (1998), with permission from Elsevier).

role of metallic glass is not that of a conventional reinforcement: it is introduced into tungsten to affect its fracture behavior and to increase the shear localization effects [45].

Composite containing 50-70 vol.% of W, the rest being the amorphous alloy Vitralloy 106a, have been prepared by Mathaudhu et al [36] by severe plastic deformation of the mixture of the corresponding powders. Namely, warm equal channel angular extrusion was used at temperatures in the super-cooled region of the metallic glass (410-420°C). This allowed fabricating composites with significant but insufficient bonding between the metallic glass and the tungsten phase. Although the metallic glass experienced viscous flow during the processing, that was not enough to ensure good bonding, because for tungsten, the temperatures used during warm equal channel angular extrusion were very low to induce diffusion. The authors suggest using lower melting temperature metals instead of tungsten in further investigations and wish to have a non-agglomerated powder of the matrix so that the interfacial area in the composite could increase, which would promote better bonding in the composite by the viscous metallic glass binder during the warm processing. There is hope that various composites can be prepared by severe plastic deformation of metallic glass/metal powder blends because the enhanced contribution of deformation processes to consolidation and densification can to some extent ease the requirement for certain values of melting temperatures of the matrix metal to make a dense composite by powder metallurgy.

In a very recent study by Lee et al [45], metallic glass-reinforced tungsten was produced using Spark Plasma Sintering of the milled powder mixture of W and a Hf-based amorphous alloy. From the known capabilities of Spark Plasma Sintering and similar electric-current assisted methods [73-75] to facilitate sintering at temperatures lower than those chosen for conventional processes, one can anticipate very promising results for difficult-to-sinter materials. The optimized sintering conditions made it possible to reach 92.8 and 94.7% relative densities in the composites with 50 and 70 vol.% of W, respectively. The sintered materials exhibited shear localization together with enhanced plasticity, which have been sought in these composites for some time [45].

In terms of possible processing routes of tungsten-metallic glass composites, there exists an opportunity to first fabricate a porous tungsten perform and then infiltrate it with a melt transforming upon cooling into a metallic glass [71-72]. As a result of such processing, metallic glass can be evenly distributed in the porous tungsten matrix (Fig.17, [71]). This methods appears very attractive for creating composites with high-melting temperature matrices, however, possible interactions of the molten multi-component alloy with the matrix metal and the wettability issues should be taken into account.

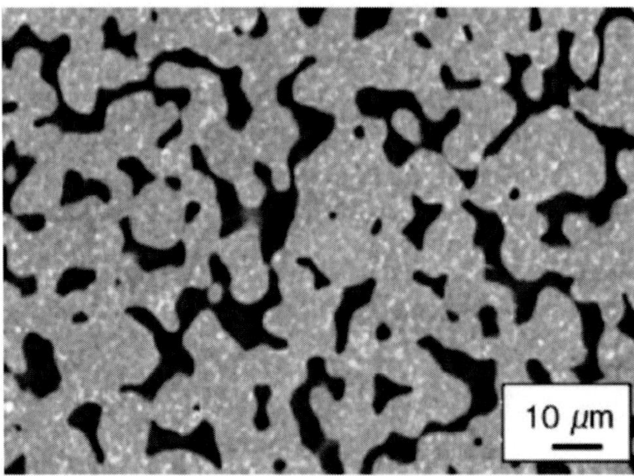

Figure 17. SEM image of the Zr-based metallic glass reinforced porous tungsten composite [71] (Reprinted from Mater.Sci.Eng. A445-446, Y.F.Xue, H.N.Cai, L.Wang, F.C.Wang, H.F.*Zhang*. Dynamic compressive deformation and failure behavior of Zr-based metallic glass reinforced porous tungsten composite, Pages 275-280, Copyright (2007), with permission from Elsevier).

4. Current Challenges and Future Perspectives

The review of the existing up to date literature shows that significant advancements have been made in the past few years in the field of metallic glass-reinforced MMCs. However, at the microstructural level, the design of the composites can be further continued by reducing the particle size of the metallic glass phase. Special attention can be directed to the following aspects: possibilities of using light-weight Al-based metallic glasses and metallic glasses with relatively high glass transition temperatures T_g and high strength. The challenge to be tackled when sintering a composite with an Al-based metallic glass will be to find a temperature window suitable to serve the purposes of full densification while preserving its glassy nature. Metallic glasses with higher T_g possess higher strength, so, the problem to solve is to find a suitable powder processing procedure to mix a relatively soft matrix alloy powder with very hard and strong metallic glass. While this problem is challenging, it is worth tackling because higher T_g metallic glasses would allow using higher melting temperature alloys as matrices.

Recent studies show that metallic glass-reinforced MMCs can give a good combination of strength and ductility in compression, which encourages further research in this area. Yield and fracture strength that are achievable in these composites can be higher or comparable with those of ceramic particle-reinforced MMCs. However, tensile properties of metallic glass-reinforced MMCs are yet to be investigated.

The interest to this class of composite materials is currently expressed by several research groups in France, Germany, USA, China, South Korea and Japan. Future developments in the research field of metallic glass, such as new compositions of amorphous alloys with increased thermal stability and high strength, can inspire the development of new composite combinations.

Conclusion

In the recent years, alternative reinforcements for metallic matrices – particles of amorphous structure – have been proposed, which created a basis for the rapid development of a new class of MMCs. The reinforcing particles are amorphous multi-component metallic alloys of high strength and 2% elastic strain limit. The attractiveness of metallic glass as a reinforcing phase lies in that it has the same metallic nature as the matrix material and can presumably have better compatibility with the matrix than conventional ceramic reinforcements in terms of chemical and physical properties.

Metallic glass-reinforced MMCs can be produced using infiltration of a molten matrix alloy into a porous preform made of metallic glass or a molten alloy that will form an amorphous phase can be infiltrated into a porous preform of a high-melting temperature matrix metal, such as W. In some specific cases, metallic glass ribbons can be pressed in between metal foils and sintered or rolled. For certain compositions, it is possible to form a crystalline metal phase in high volume contents during solidification while keeping the other phase (in minor volume content) amorphous. However, the most versatile method to produce amorphous phase-reinforced MMCs is powder metallurgy.

Above its glass transition temperature T_g, in the super-cooled liquid region, metallic glass particles become soft and act as a viscous binder and porosity remover in the composite. After compaction to full density and cooling to below T_g and down to room temperature, the same glassy particles become the hard reinforcement phase. Since metallic glass crystallizes above a certain temperature T_x, it is particularly suitable for low melting temperature alloys, such as those based on aluminum and magnesium. The composite materials should be designed in such a way that the chosen sintering temperature of the composite powders is within the super-cooled liquid region of the glass while being very close to the solidus temperature of the matrix alloy in order to provide conditions for successful consolidation and pore elimination between the sintering composite particles.

At present, the most popular metal matrices to be reinforced with metallic glass particles are Al and its alloys. Research on Mg alloy matrix composites reinforced with metallic glass has also been started. Higher melting temperature metals, such as Ti, attract attention and are among future candidates for the matrix materials.

REFERENCES

[1] S.C.Tjong, Z.Y.Ma. *Microstructural and mechanical characteristics of in situ metal matrix composites.* Mater. Sci.Eng. 29 (2000) 49-113.

[2] J.W.Kaczmar, K.Pietrzak, W.Wlosinski. *The production and application of metal matrix composite materials.* J. Mater. Proc. Tech. 106 (2000) 58-67.

[3] J.M.Torralba, C.E.da Costa, F.Velasco. *P/M aluminum matrix composites: an overview.* J.Mater.Proc. Tech. 133 (2003) 203-206.

[4] D.B.Miracle. *Metal matrix composites – From science to technological significance.* Comp. Sci. Tech. 65 (2005) 2526–2540.

[5] S.C.Tjong. *Novel nanoparticle-reinforced metal matrix composites with enhanced mechanical properties.* Adv. Eng. Mater. 9 (8) (2007) 639-652.

[6] A.Mortensen, J.Llorca. *Metal Matrix Composites.* Annu. Rev. Mater. Res. 40 (2010) 243–270.

[7] A.Slipenyuk, V.Kuprin, Y.Milman, V.Goncharuk, J.Eckert. *Properties of P/M processed particle reinforced metal matrix composites specified by reinforcement concentration and matrix-to-reinforcement particle size ratio.* Acta Mater. 54(2006) 157-166.

[8] M.L.Pines, H.A.Bruck. *Pressureless sintering of particle-reinforced metal–ceramic composites for functionally graded materials: Part I. Porosity reduction models.* Acta Mater. 54 (2006) 1457-1465.

[9] S.J.Cytron. *Method of making metallic glass-metal matrix composites.* US Patent 4,562,951 (1986).

[10] A.K.Jha, G.S.Upadhyaya, P.K.Rohatgi. *Properties of composites of 2014 aluminum alloy with Ni-Mo-based metallic glass particles.* J.Mater.Sci. 21 (1986) 1502-1508.

[11] M.T.Stawovy, A.O.Aning. *Processing of amorphous Fe-W reinforced Fe matrix composites.* Mater. Sci.Eng. A256 (1998) 138-143.

[12] A.Inoue. *Stabilization of metallic supercooled liquid and bulk amorphous alloys.* Acta Mater. 2000; 48 (1): 279-306.

[13] A.Inoue, B.Shen, H.Koshiba, H.Kato, A.R.Yavari. *Cobalt-based bulk glassy alloy with ultrahigh strength and soft magnetic properties.* Nature Mater. 2 (10) (2003) 661-663.

[14] M.F.Ashby, A.L.Greer. *Metallic glasses as structural materials.* Scripta Mater. 54 (2006) 321-326.

[15] A.R.Yavari, J.J.Lewandowski, J.Eckert. *Mechanical Properties of bulk metallic glasses.* MRS Bull. 32 (2007) 635-638.

[16] A.Inoue, N.Nishiyama. *New bulk metallic glasses for applications as magnetic-sensing, chemical and structural materials.* MRS Bull. 32 (8) (2007) 651-658.

[17] D.V.Dudina, K.Georgarakis, Y. Li, M.Aljerf, A.LeMoulec, A.R.Yavari, A.Inoue. *A magnesium alloy matrix composite reinforced with metallic glass.* Comp. Sci. Tech. 69 (15-16) (2009) 2734-2736.

[18] C.A. Schuh, T.C.Hufnagel, U.Ramamurty. *Mechanical behavior of amorphous alloys.* Acta Mater. 55 (2007) 4067–4109.

[19] K.Georgarakis, M.Aljerf, Y.Li, A.LeMoulec, F.Charlot, A.R.Yavari, K. Chornokhvostenko, E.Tabachnikova, G.A.Evangelakis, D.B.Miracle, A.L.Greer, T.Zhang. *Shear band melting and serrated flow in metallic glasses.* Appl. Phys. Lett. 93 (3) (2008) 031907 1 – 3.

[20] M.Chen. *Mechanical Behavior of metallic glasses: microscopic understanding of strength and ductility.* Annu. Rev. Mater. Res. 38 (2008) 445–469.

[21] G.He, J.Eckert, W.Löser, L.Schultz. *Novel Ti-base nanostructure–dendrite composite with enhanced plasticity.* Nature Mater. 2 (2003) 33-37.

[22] D.C. Hofmann, J.-Y.Suh, A.Wiest, G.Duan, M.Lind, M.D.Demetriou, W.L.Johnson. *Designing metallic glass matrix composites with high toughness and tensile ductility.* Nature 451 (2008) 1085-1090.

[23] D.C.Hofmann, J.-Y. Suh, A.Wiest, M.-L.Lind, M.D.Demetriou, W.L.Johnson. *Development of tough, low-density titanium-based bulk metallic glass matrix composites with tensile ductility.* Proc. Nat.Acad.Sci. 105 (51) (2008) 20136-20140.

[24] F.Q.Guo, S.J.Poon, G.J.Shiflet. *Networking amorphous phase reinforced titanium composites which show tensile plasticity.* Phil.Mag.Lett. 88 (2008) 615-622.

[25] A. Inoue, W. Zhang, T.Tsurui, A.R.Yavari, A.L.Greer. *Unusual room-temperature compressive plasticity in nanocrystal-toughened bulk copper-zirconium glass.* Phil. Mag.Lett. 85 (5) (2005) 221-229.

[26] A.M. Jorge Jr., M.J.E. Rodrigues, M.F. de Oliveira, A.R.Yavari, C.Bolfarini, C.S.Kiminami, W.J. Botta. *Partial crystallisation and mechanical properties of bulk metallic glasses.* J. Metastable Nanocryst. Mater. 20-21 (2004) 71-76.

[27] K. Hajlaoui, A.R.Yavari, A.LeMoulec, W.J. Botta, F.G.Vaughan, J. Das, A.L.Greer, Å.Kvick. *Plasticity induced by nanoparticle dispersions in bulk metallic glasses.* J. Non-Cryst. Solids 353 (2007) 327–331.

[28] K. Hajlaoui, A.R. Yavari, J. Das, G.Vaughan. *Ductilization of BMGs by optimization of nanoparticle dispersion.* J. Alloys Comp. 434–435 (2007) 6–9.

[29] M.H.Lee, J.H.Kim, J.S.Park, J.C.Kim, W.T.Kim, D.H.Kim. *Fabrication of Ni–Nb–Ta metallic glass reinforced Al-based alloy matrix composites by infiltration casting process.* Scripta Mater. 50 (2004) 1367-1371.

[30] K.Hono, Y.Zhang, T.Sakurai, A.Inoue. *Microstructure of a rapidly solidified Al-4V-2Fe ultrahigh strength aluminum alloy.* Mater. Sci.Eng. A 250 (1998) 152-157.

[31] A.Inoue. *Amorphous, nanoquasicrystalline and nanocrystalline alloys in Al-based systems.* Progress in Mater. Sci. 43 (1998) 365-520.

[32] M.H.Lee, J.S.Park, J.H.Kim, W.T.Kim, D.H.Kim. *Synthesis of bulk amorphous alloy and composites by warm rolling process*. Mater.Lett. 59 (2005) 1042-1045.

[33] J.Nagy, M.Balog, K.Izdinsky, F.Simancik, P.Svec, D.Janickovic. *High strength potential of aluminum nanocomposites reinforced with nonperiodical phases*. Intl. J. Mater. Prod. Tech. 23 (2005) 79-90.

[34] C.A.Wensley, A.O.Aning, J.Schulz, S.Kampe. *Processing of amorphous Ni-W reinforced Ni matrix composites*. Adv.Mater.Proc. 163 (2005) 48.

[35] P.Yu, K.B.Kim, J.Das, F.Baier, W.Xu, J.Eckert. *Fabrication and mechanical properties of Ni-Nb metallic glass particle-reinforced Al-based metal matrix composite*. Scripta Mater. 54 (2006) 1445-1450.

[36] S.N.Mathaudhu, K.T.Hartwig, I.Karaman. *Consolidation of blended powders by severe plastic deformation to form amorphous metal matrix composites*. J. Non-Cryst. Solids 353 (2007) 185-193.

[37] A.Samanta, H.J.Fecht, I.Manna, P.P.Chattopadhyay. *Development of amorphous phase dispersed Al-rich composites by rolling of mechanically alloyed amorphous Al-Ni-Ti powders with pure Al*. Mater. Chem. Phys. 104 (2007) 434-438.

[38] P.Yu, L.C.Zhang, W.Y.Zhang, J.Das, K.B.Kim, J.Eckert. *Interfacial reaction during the fabrication of* $Ni_{60}Nb_{40}$ *metallic glass particles-reinforced Al based MMCs*. Mater. Sci. Eng. A444 (2007) 206-213.

[39] P.Yu, S. Venkataraman, J.Das, L.C.Zhang, W.Zhang, J.Eckert. *Effect of high pressure during the fabrication on the thermal and mechanical properties of amorphous* $Ni_{60}Nb_{40}$ *particle-reinforced Al-based metal matrix composites*. J.Mater.Res. 22 (2007) 1168-1173.

[40] S. Scudino, K.B.Surreddi, S.Sager, M.Sakaliyska, J.S.Kim, W.Löser, J.Eckert. *Production and mechanical properties of metallic glass-reinforced Al-based metal matrix composites*. J. Mater. Sci. 43 (2008) 4518–4526.

[41] J.Eckert, M.Calin, P.Yu, L.C.Zhang, S.Scudino, C.Duhamel. *Al-based alloys containing amorphous and nanostructured phases*. Rev.Adv.Mater.Sci. 18 (2008) 189-172.

[42] S. Scudino, G.Liu, K.G.Prashanth, B.Bartusch, K.B.Surreddi, B.S.Murty, J.Eckert. *Mechanical properties of Al-based metal matrix composites reinforced with Zr-based glassy particles produced by powder metallurgy*. Acta Mater. 57 (2009) 2029-2039.

[43] X.L.Zhang, J.X.Wang, Y.X.Sun, J.C.Liu. *Metallic glass particle reinforced Al-based and (Al-Ni)-based metal matrix composites*. Comb. Expl. Shock Waves 45 (2009) 230-235.

[44] D. V. Dudina, K. Georgarakis, Y. Li, M. Aljerf, M. Braccini, A. R. Yavari, A. Inoue. *Cu-based metallic glass particle additions to significantly improve overall compressive properties of an Al alloy*. Composites Part A 41 (2010) 1551-1557.

[45] M.H.Lee, J.K.Lee, K.B.Kim, D.J.Sordelet, J.Eckert, J.C.Bae. *Mechanical behavior of metallic glass reinforced nanostructured tungsten composites synthesized by spark plasma sintering*. Intermetallics (2010) in press.

[46] A.R.Yavari, W.J. Botta Filho, C.A.D.Rodrigues, K.R.Cardoso, R.Valiev. *Nanostructured bulk* $Al_{90}Fe_5Nd_5$ *prepared by cold consolidation of gas atomised powder using severe plastic deformation*. Scripta Mater. 46 (2002) 711-716.

[47] J.Ye, B.Q.Han, J.M.Schoenung. *Mechanical behaviour of an Al–matrix composite reinforced with nanocrystalline Al-coated* B_4C *particulates*. Phil. Mag. Lett. 86 (11) (2006) 721–732.

[48] S.K.Thakur, M.Gupta. *Improving mechanical performance of Al by using Ti as reinforcement.* Composites Part A 38 (3) (2007) 1010-1018.

[49] A.Inoue, H.M.Kimura, T.Zhang. *High-strength aluminum- and zirconium-based alloys containing nanoquasicrystalline particles.* Mater.Sci.Eng. 294-296 (2000) 727-735.

[50] F.Audebert, F.Prima, M.Galano, M.Tomut, P.J.Warren, I.C.Stone, B.Cantor. *Structural characterization and mechanical properties of nanocomposite Al-based alloys.* Mater.Trans. 43 (8) (2002) 2017-2025.

[51] M.Galano, F.Audebert, B.Cantor, I.Stone. *Structural characterization and stability of new nanoquasicrystalline Al-based alloys.* Mater.Sci.Eng A 375-377 (2004) 1206-1211.

[52] M.Galano, F.Audebert, I.C.Stone, B.Cantor. *Effect of Nb on nanoquasicrystalline Al-based alloys.* Phil. Mag. Lett. 88 (2008) 269–278.

[53] Y.Li, K.Georgarakis, S.Pang, F.Charlot, A.LeMoulec, S.B.Profeta, T.Zhang, A.R.Yavari. *Chill-zone aluminum alloys with GPa strength and good plasticity.* J. Mater. Res. 24 (2009) 1513-1521.

[54] Y.Li, Georgarakis K, Pang S, Antonowicz J, Charlot F, LeMoulec A, Zhang T, Yavari AR. *AlNiY chill-zone alloys with good mechanical properties.* J. Alloys Comp. 477 (2009) 346-349.

[55] Z.Zhang, B.Q.Han, D.Witkin, L.Ajdelsztajn, E.J.Lavernia. *Synthesis of nanocrystalline aluminum matrix composites reinforced with in situ Al-Ni-La amorphous particles.* Scripta Mater. 54 (2006) 869-874.

[56] W.J. Botta Filho, J.B. Fogagnolo, C.A.D. Rodrigues, C.S. Kiminami, C. Bolfarini, A.R. Yavari. *Consolidation of partially amorphous aluminum-alloy powders by severe plastic deformation.* Mater. Sci. Eng. A 375–377 (2004) 936–941.

[57] K.K.Chawla. *Composite Materials: Science and Engineering.* 2nd edition, Springer Science+Business Media LLC, New York, USA, 1998.

[58] Z.H.Tan, B.J.Pang, D.T.Qin, J.Y.Shi, B.Z.Gai. *The compressive properties of 2024Al matrix composites reinforced with high content SiC particles at various strain rates.* Mater.Sci.Eng.A 2008; 489 (1-2): 302-309.

[59] R.Bauri, M.K.Surappa. *Processing and compressive strength of Al–Li–SiC$_p$ composites fabricated by a compound billet technique.* J.Mater.Proc.Tech. 2009; 209 (4): 2077-2084.

[60] J.Cai, Y.Chen, V.F.Nesterenko, M.A.Meyers. *Effect of strain rate on the compressive mechanical properties of aluminum alloy matrix composite filled with discontinuous carbon fibers.* Mater. Sci. Eng. A 485 (2008) 681-689.

[61] R.M.German. *Powder Metallurgy and Particulate Materials Processing,* Metal Powder Industries Federation, USA, 2005.

[62] H.Z.Ye, X.Y.Liu. *Review of recent studies in magnesium matrix composites.* J.Mater.Sci. 39 (2004) 6153-6171.

[63] Y.L.Xi, D.L.Chai, W.X.Zhang, J.E.Zhou. *Titanium alloy reinforced magnesium matrix composite with improved mechanical properties.* Scripta Mater. 54 (2006) 19-23.

[64] S.F.Hassan, M.Gupta. *Development of ductile magnesium composite materials using titanium as reinforcement.* J.Alloys Comp. 345 (2002) 246-251.

[65] W.W.Leong, M.Gupta. *Enhancing thermal stability, modulus and ductility of magnesium using molybdenum as reinforcement.* Adv. Eng. Mater. 7 (2005) 250-256.

[66] S.F.Hassan, M.Gupta. *Development of ductile magnesium composite materials using titanium as reinforcement.* J. Alloys Comp. 345 (2002) 246-251.

[67] S.F.Hassan , M.Gupta. *Development of high strength magnesium based composites using elemental nickel particulates as reinforcement.* J. Mater. Sci. 37 (2002) 2467-2474.

[68] Z.Trojanova, V.Gartnerova, A.Jager, A.Namesny,M.Chalupova, P.Palcek, P.Lukac. *Mechanical and fracture properties of an AZ91 Magnesium alloy reinforced by Si and SiC particles Comp.* Sci. Tech. 69 (2009) 2256-2264.

[69] A.K.Khanra, H.C.Jung, K.S.Hong, K.S.Shin. *Comparative property study on extruded Mg–HAP and ZM61–HAP composites.* Mater. Sci. Eng. A 527 (2010) 6283–6288.

[70] Q.B.Nguyen, M. Gupta. *Enhancing compressive response of AZ31B using nano-Al_2O_3 and copper additions.* J. Alloys Comp. 490 (2010) 382–387.

[71] Y.F.Xue, H.N.Cai, L.Wang, F.C.Wang, H.F.Zhang. *Dynamic compressive deformation and failure behavior of Zr-based metallic glass reinforced porous tungsten composite.* Mater.Sci.Eng. A445-446 (2007) 275-280.

[72] Y.F. Xue, L. Wang, H.W. Cheng, F.C. Wang, H.F. Zhang. *Shear band formation and mechanical properties of $Zr_{38}Ti_{17}Cu_{10.5}Co_{12}Be_{22.5}$ bulk metallic glass/porous tungsten phase composite by hydrostatic extrusion.* Mater. Sci. Eng. A 527 (2010) 5909–5914.

[73] Z.A.Munir, U.Anselmi-Tamburini, M.Ohyanagi. *The effect of electric field and pressure on the synthesis and consolidation of materials: A review of the spark plasma sintering method.* J. Mate.r Sci. 41(3) (2006) 763-777.

[74] D.M.Hulbert, A.Anders, D.V.Dudina, J.Andersson, D.Jiang, C.Unuvar, U.Anselmi-Tamburini, E.J.Lavernia, A.K.Mukherjee. *The absence of plasma in "Spark Plasma Sintering".* J. Appl. Phys. 104 (2008) 033305-1 - 033305-7.

[75] E.A.Olevsky, S.Kandukuri, L.Froyen. *Consolidation enhancement in spark-plasma sintering: Impact of high heating rates.* J. Appl. Phys. 102 (2007) 114913-1 - 114913-12.

In: Metal Matrix Composites
Editor: J. Paulo Davim

ISBN: 978-1-61209-771-8

Chapter 2

"MAXMET"S: A NEW CLASS OF METAL MATRIX COMPOSITES REINFORCED WITH MAX PHASES

Shahram Amini[1*] and Michel W. Barsoum[2]
[1]National Hypersonic Science Center, Materials Department,
University of California, Santa Barbara, California, US
Department of Materials Science and Engineering,
[2]Drexel University, Philadelphia, Pennsylvania, US

ABSTRACT

The MAX phases and hexagonal metals, among many other plastically anisotropic solids with c/a ratios > 1.5, have been recently classified as kinking nonlinear elastic (KNE) solids – the signature of which is the formation of fully reversible, hysteretic stress-strain loops during cyclic loadings. In this review, recent progress in a novel class of Mg matrix composites reinforced with Ti_2AlC, what is known as MAXMET, is reviewed. The Mg grains that constitute the matrix of these composites are exceptionally stable, a fact attributable to nanocrystalline nature of the Mg-matrix. This review shows that the MAXMET composites are all KNE solids and their KNE behavior is a strong function of texture, and characterized by the formation of fully reversible hysteretic stress-strain loops under uniaxial cyclic compression.

INTRODUCTION

The MAX phases are layered hexagonal solids, with two formula units per unit cell, in which near close-packed layers of M are interleaved with layers of pure A-group elements, with the X-atoms filling the octahedral sites between the M layers. At this time it is fairly well established that these phases have an unusual and sometimes unique combination of properties. They are excellent electrical and thermal conductors, thermal shock resistant and damage tolerant [1-6]. Despite being elastically quite stiff, they are all readily machinable

*Corresponding author, E-mail: shahram@engineering.ucsb.edu

with nothing more sophisticated than a manual hacksaw [3, 4]. Moreover, some of them are fatigue, creep and oxidation resistant [5-17].

More recently [18-23], the MAX phases were classified as kinking nonlinear elastic (KNE) solids because they deform primarily by kinking, and the formation of kink bands. Kinking – a mechanism first reported in crystalline solids by Orowan in single crystals of Cd loaded parallel to the basal planes [24] – has also been identified as the physical origin of the hysteretic, nonlinear elastic behavior exhibited by these solids. Kink band formation is the key mechanism without which the deformation of KNE solids, in general, and MAX phases in particular, cannot be understood. When cyclically loaded, KNE solids form fully reversible, hysteretic stress-strain loops (Figure 1). This full reversibility has been attributed to the formation of IKBs that are comprised of multiple, parallel dislocation loops, whose shape ensures that when the load is removed they shrink significantly or are annihilated altogether (Figure 2).

Recently, it was established that a number of seemingly unrelated solids such as the MAX phases [23, 25, 26], graphite [21], mica [27], sapphire [28], ZnO [29], GaN [30], $LiNbO_3$ [31] and the hexagonal metals (Mg, Co, Ti, Zn etc.) [32-36], among many others, are KNE solids. Most recently, and as will be shown in this chapter, MAX-reinforced metal-matrix composites – what can be labeled MAXMETs – are KNE solids too [37, 38]. The signature of KNE solids is the formation of fully reversible, hysteretic stress-strain loops (Figure 1) during cyclic loadings [23, 39]. The various parameters needed to describe the stress-strain loops are as follows: measured stress (σ), nonlinear strain (ε_{NL}), linear strain (ε_{LE}), and dissipated energy per unit volume per cycle (W_d). Full reversibility of such loops has been attributed to the formation of incipient kink bands (IKBs) that are comprised of multiple, parallel dislocation loops (Figure 2), whose shape ensures that when the load is removed they shrink significantly or are annihilated altogether [40]. In other words, solids that are highly plastically anisotropic deform at least initially by the formation of dislocation-based IKBs [41].

In addition to being fully reversible, the loading-unloading stress-strain curves of KNE solids are strain-rate independent and highly reproducible. Their shape, and the energy dissipated per unit volume in each cycle, W_d, are strongly influenced by grain size, with the former being significantly larger in coarse-grained solids [23]. At room temperature, some KNE solids such as Ti_3SiC_2 can be compressed to stresses as high as 1 GPa and fully recover upon removal of the load, while dissipating $\approx$ 25% of the mechanical energy [23, 42]. Recently, we developed a microscale model for this deformation behavior [34] that is based on the work of Frank and Stroh, F&S, [40]. Detailed versions of this model can be found in Refs. [18, 26, 34, 35, 43-48]. A simplified version of the model is presented below.

Kinking nonlinear elasticity was first documented in Ti_3SiC_2 (Figure 3), a founding member of the MAX phases [18-23, 49].

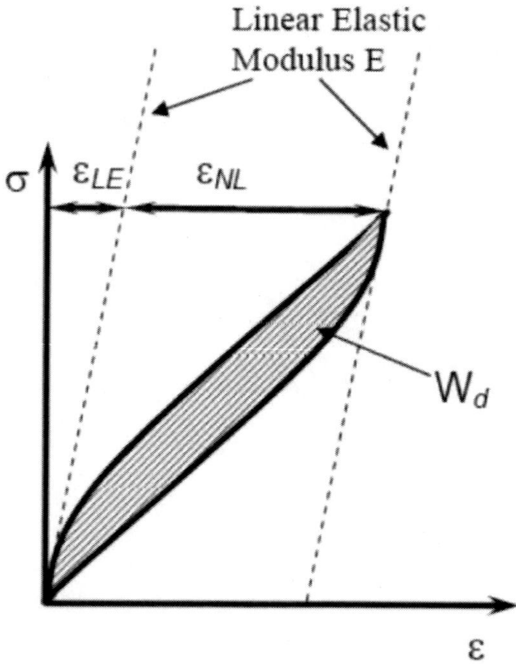

Figure 1. Schematic of a typical stress-strain curve for a KNE solid. The various parameters needed to describe the curve are labeled.

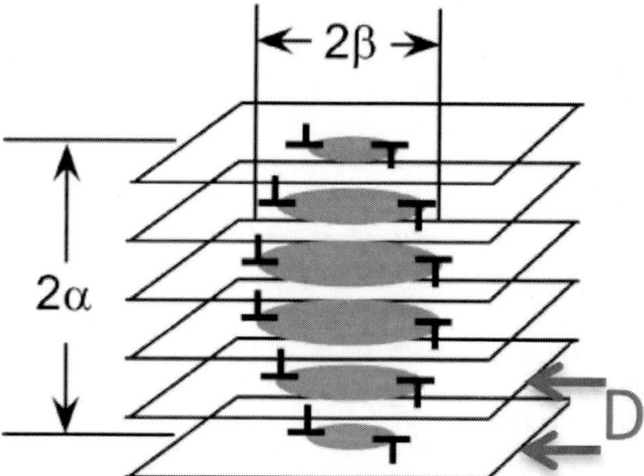

Figure 2. Schematic of an IKB with length 2α and diameter 2β. D is the distance between the horizontal dislocation loops.

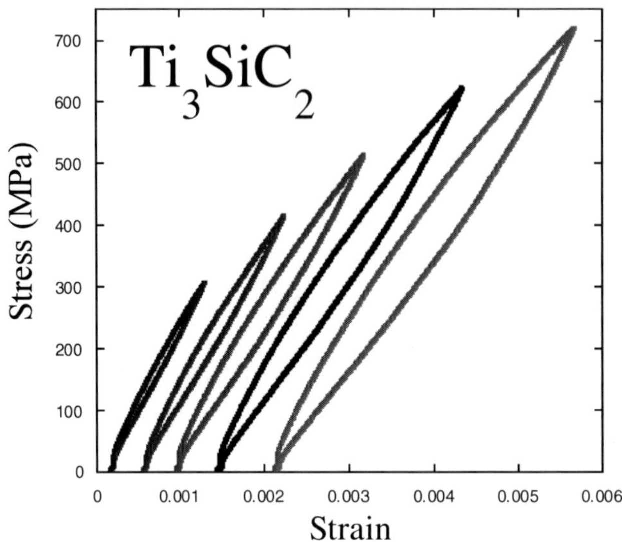

Figure 3. Typical stress-strain curves of Ti_3SiC_2; the loops are shifted horizontally for clarity.

Recently, Zhou et al. reported on the mechanical response of fully dense and 10 vol. % porous Ti_2AlC bulk samples to cyclic compressive loadings [26]. In that work, it was shown that the porous material dissipates more energy on an absolute scale, which is compelling evidence that we are dealing with a kink-based phenomenon as opposed to one that is dependent on the volume of the material, such as dislocation pileups.

More recently [25], it was also shown that Cr_2GeC samples compressively loaded from 300 MPa to ~ 570 MPa exhibited nonlinear, fully reversible, reproducible, hysteretic loops that dissipated ~ 20 % of the mechanical energy, due to the formation and annihilation of IKBs [45].

Ever since the development of the Volkswagen Beetle magnesium, Mg, engines and transmission in 1946 until the Mercedes-Benz 300 SLR Le Mans disaster in 1955, Mg parts had been seen as candidates for automotive components [50]. More importantly, however, Mg, and its alloys have also been recently used extensively in various industries – such as automotive – due to their lightweight, good castability and machinability [51-53]. Mg is also well known for its high damping capabilities [51, 54]. The reasons for this high damping had to date not been well understood. Most recently, it was shown by Zhou et al. that the high damping can be traced to the formation of IKBs [34, 35]. In other words, they showed that Mg – and other hexagonal metals, including Ti, Co and Zn – can be classified as KNE solids.

Compared to other structural metals, Mg alloys have relatively low strength, especially at elevated temperatures that limits their applications to temperatures lower than ≈ 120 °C. The need for high-performance and lightweight materials for some demanding applications has led to the development of Mg-matrix composites [51, 55]. Despite their advantages, a major drawback of Mg matrix composites is their relative high cost of fabrication. It follows that cost-effective processing would expand their applications [51]. To manufacture composites with optimum properties, the manufacturing process must assure a uniform distribution of the reinforcing phase in the matrix. A variety of Mg-matrix composites have been fabricated through powder metallurgy [56-58]. Stir casting has been used for manufacturing composites, with up to 30% vol. fraction of reinforcement. Further extrusion to reduce porosity, refine the microstructure, and homogenize the distribution of the reinforcement has also been used [51, 59-62]. Squeeze infiltration [51, 52, 63-68] and spontaneous infiltration [51, 69-71], have also been used. The mechanical properties of some of these composites are summarized in Table 1.

Among all feasible fabrication techniques, the advantages of melt infiltration, MI, include the capability of incorporating a relatively high volume fraction of reinforcement and the fabrication of composites with matrix alloys and reinforcement systems that are otherwise not easily fabricated by other techniques. The MI technique not only resulted in homogenous microstructures but, more importantly, is a low cost technique that can be readily scaled up.

Given the unique properties of Ti_2AlC [73], its high damping [26], and the advantages of Mg [35], it was postulated that Mg-Ti_2AlC composites should result in solids that are not only machinable, stiff and light, but would also exhibit exceptional damping capabilities. Some of the objectives of this chapter are as follows: To report on the mechanical response of the aforementioned composites under microhardness indentation and compressive loadings; to apply the KNE microscale model to analyze the cyclic compressive stress-strain curves

Table 1. Mechanical properties of select Mg matrix composites fabricated with different techniques and various volume fractions of reinforcement (E is Young's Modulus; UTS is Ultimate Tensile Strength)

Matrix	Reinforcement	Fabrication Technique	E (GPa)	UTS (MPa)	Ref.
AZ91*	-	As-cast	40	150-200	[56, 63]
AZ80**	50 vol.% SiC	Pressure infiltration	103	550	[68]
AZ91	20 vol.% alumina	Infiltration	70	240-290	[63]
AZ91	10 vol.% SiC	Mechanical alloying	50	150	[57]
AZ91	10 vol.% SiC	Powder metallurgy	45-50	135	[56, 57]
Mg	30 vol.% SiC	Powder metallurgy	59	250	[61]
Mg	30 vol.% SiC	Stir cast and extruded	60	258	[61]
Mg	50 vol.% SiC	Hot pressing (HPing)	120±20	-	[43, 44]
Mg	50 vol.% Ti_3SiC_2	HPing	75±5	-	[43, 44]
Mg	50 vol.% Ti_2AlC	Melt infiltration (MI)	72±6	350±40	[43, 44]

*(Cast Mg alloy; 9 % Al, 1 % Zn and 0.2 % Mn) ** (Cast and wrought Mg alloy; 8 % Al, 0.5 % Zn and 0.2 % Mn) [72]

obtained. Lastly, because the MAX phases are layered hexagonal solids that deform by kinking, it was postulated that their response to cyclic loadings would depend on the orientation of the basal planes relative to the loading direction. The effect of texture was thus investigated.

In general there has not been much work on MAX-metal composites – what can be labeled MAXMETs. Recently, Gupta et al. [74] fabricated Ta2AlC and Cr2AlC Ag-based composites. These composites are new solid lubricant materials for use over a wide temperature range against Ni-based superalloys and alumina. There are also a few papers in the literature on Ti3SiC2-Cu composites [75-77] and Ti3SiC2-SiC and TiC composites [78, 79]. Warm compaction, which is a simple and economical forming process to prepare high density powder metallurgy materials, was employed by Ngai et. al [76, 77] to fabricate Ti3SiC2 particulate reinforced Cu matrix composite with high strength, high electrical conductivity and good tribological behaviors. Ti3SiC2-Cu composites, with 1.25, 2.5 and 5 wt. % Ti3SiC2 were prepared by compacting powder with a pressure of 700 MPa at 145°C followed by sintering at 1000 °C. Their density, electrical conductivity and UTS decrease with increase in particulate concentration, while the hardness increases with the increase in particulate concentration. A small addition of Ti3SiC2 particulate increased the hardness of the composite without losing much of the electrical conductivity. The composite containing 1.25 wt. % Ti3SiC2 has an UTS of 158 MPa and an electrical resistivity of 3.91×10-8 Ωm [77]. Cu-Ti3SiC2 composite powders have also been fabricated using an electroless plating technique [75].

Wu et al. [80] reported on Ti2SnC dispersion-strengthened Cu matrix composites fabricated by hot-pressing (HPing). The change of microstructure, mechanical properties, and electrical resistivity as a function of Ti2SnC volume fraction were studied. Their results demonstrated that the grain size of Cu decreased pronouncedly from 27 μm in bulk Cu to ~ 1 μm by incorporating Ti2SnC particulates, and the strengthening effect was significant. Improvements in yield strength of up to four times that of pure Cu were found in Cu-1 vol% Ti2SnC, however, the conductivity of the composite was 85.6% of pure Cu. The high strength and low electrical resistivity of Cu-Ti2SnC composites indicated that Ti2SnC is a promising reinforcement for Cu.

For many years the use of Mg alloys as the matrix phases in metal matrix composites (MMCs) has been of interest as alternatives to Al-based composites for advanced structural applications and for components in engines, with the advantage of high specific strength and stiffness. Mg MMCs have mechanical properties basically similar to those of Al MMCs and can be used for similar lightweight structural and functional parts. The main advantage of Mg MMCs in comparison to Al MMCs is weight savings of about 20 to 25 %. In general, they also provide better machinability in comparison to Al MMCs [68, 81, 82]. Although cast Mg alloys and their composites have dominated the market, interest in the use of wrought Mg alloys and composites with greater strength and ductility is continuously growing [83].

Early development of Mg MMCs concentrated on continuous-fiber reinforced composites, especially graphite fibers due to the low thermal expansion of the composites fabricated. More recently, discontinuously reinforced composites have combined alumina, boron carbide and silicon carbide in the form of whiskers, particles and short fibers with a variety of Mg alloys through different manufacturing processes such as rheo-, compo-, stir- and squeeze-casting, and various MI processes, as well as, powder metallurgy [68].

Due to its extensive fluid-flow capabilities and lower melting temperature than Al, Mg is an ideal material for processing via the molten or semi-solid metal processing. Among all, MI processes have numerous advantages, mainly due to the reactivity and low viscosity of Mg that speeds up the production of Mg MMCs. Another important advantage is that selectively reinforced components, within which the metal is reinforced only where needed, can be produced [84].

Two broad variants of infiltration processes have been devised for the fabrication of Mg MMCs: (1) pressureless infiltration (i.e. spontaneously, in the absence of an external pressure) which can be used when good wetting conditions exist, and, (2) pressure infiltration, in which pressure is required to drive the molten metal into the preform. Pressureless infiltration is simple, cost-effective and can result in near-net shape parts with a more homogenous distribution of particulate reinforcements. Easy tailoring of reinforcement volume fraction between 35-70 vol. % or higher is also possible [68, 84].

Spontaneous MI of Mg-based alloys into preforms is usually performed in an oxygen-free, nitrogen-containing atmosphere, where in the initial stages of the processing, Mg reacts with nitrogen to form thin Mg_3N_2 coatings on the reinforcement surfaces, which in turn renders the system wettable and eliminates the need for pressure. The MI can also be carried out in vacuum [68, 84]. We show here that composites can be fabricated successfully in a vacuum atmosphere.

To manufacture composites with optimum properties, the manufacturing process has to assure a uniform distribution of the reinforcing phase in the matrix. Both techniques namely melt infiltration (MI) and HPing (HP), result in a homogenous microstructure, with a uniform distribution of the reinforcement. This is the first work on the fabrication and characterization of Mg-Ti_2AlC composites.

KNE MODEL

Recently, we developed a microscale model for this deformation behavior [34] that is based on the work of Frank and Stroh, F&S, [40]. In what follows, a simplified version is presented. F&S [40] considered an elliptic kink band, KB, with length, 2α, and width, 2β, such that α >> β (Figure 2) and showed that the remote shear stress, τ, needed to render such a subcritical KB unstable is given by:

$$\tau > \tau_t \approx \frac{\sigma_t}{M} \approx \sqrt{\frac{4G^2 b\gamma_c}{2\alpha\pi^2} \ln(\frac{b}{\gamma_c w})} \tag{1}$$

where τ_t, and σ_t are the remote critical shear and axial stresses; M is the Taylor factor relating them; G is the shear modulus and b is the Burgers vector; w is related to the dislocation core width [40]. In the MAX phase work to date, we equated the grain dimension along the [0001] direction – i.e. normal to the direction of easy slip – with 2α [18, 85]. If σ_t – that is an experimentally determinable threshold stress – is known, then 2α can be estimated from Eq. 1; γ_c is critical kinking angle calculated assuming [35, 40, 86]

$$\gamma_c = \frac{b}{D} \approx \frac{3\sqrt{3}(1-\nu)}{8\pi e}(\frac{b}{w}) \tag{2}$$

where ν is Poisson's ratio, and D is the distance between dislocation loops along 2α. Because an IKB consists of multiple parallel dislocation loops, as a first approximation, we assume each loop to be comprised of two edge, and two screw dislocation segments with lengths, $2\beta_x$ and $2\beta_y$, respectively. The latter are related to the applied stress, σ and 2α assuming [34, 40]:

$$2\beta_x \approx \frac{2\alpha(1-\nu)}{G\gamma_c}\frac{\sigma}{M} \quad \text{and} \quad 2\beta_y \approx \frac{2\alpha}{G\gamma_c}\frac{\sigma}{M} \tag{3}$$

The formation of an IKB can be divided into two stages: nucleation and growth [32]. Since the former is not well understood, the model only considers IKB growth from $2\beta_{xc}$ and $2\beta_{yc}$ to $2\beta_x$ and $2\beta_y$, respectively. The dislocation segment lengths of an IKB nucleus, β_{xc} and $2\beta_{yc}$, are presumed to pre-exist, or are nucleated during pre-straining. The values of $2\beta_{xc}$ and $2\beta_{yc}$ are estimated from Eq. 3, assuming $\sigma = \sigma_t$, where the latter is experimentally obtained (see below).

It follows that for $\sigma > \sigma_t$, the IKB nuclei grow and the IKB-induced axial strain resulting from their growth is assumed to be given by [32]:

$$\varepsilon_{IKB} = \frac{\Delta V N_k \gamma_c}{k_1} = \frac{N_k \gamma_c 4\pi\alpha(\beta_x\beta_y - \beta_{c,x}\beta_{c,y})}{3k_1} = \frac{4\pi(1-\nu)N_k\alpha^3}{3k_1 G^2\gamma_c M^2}(\sigma^2 - \sigma_t^2) = m_1(\sigma^2 - \sigma_t^2) \tag{4}$$

where m_1 is the coefficient before the term in brackets in the fourth term; N_k is the number of IKBs per unit volume; ΔV is the volume change due to one IKB as the stress is increased from σ_t to σ. It follows that the product $V \times N_k = v_f$, is the volume fraction of the material that is kinked. The factor k_1 relates the volumetric strain due to the IKBs to the axial strain along the loading direction, assumed to be 2 [87]. We note is passing that Reed-Hill *et al.* [88] also assumed $k_1 = 2$ when modelling the growth of twins in Zr.

The growth of the IKBs from $\beta_{i,c}$ to β_i leads to W_d (shaded area in Figure 1) given by [32]:

$$W_d = \frac{4\Omega\pi N_k\alpha}{D}(\beta_x\beta_y - \beta_{xc}\beta_{yc}) = \frac{4\pi(1-\nu)N_k\alpha^3}{G^2\gamma_c M^2}\frac{\Omega}{b}(\sigma^2 - \sigma_t^2) = m_2(\sigma^2 - \sigma_t^2) \tag{5}$$

where Ω is the energy dissipated by a dislocation line sweeping a unit area. Thus, Ω/b should be proportional, if not equal, to the critical resolved shear stress, CRSS, of an IKB dislocation loop. Previously we have shown that to be the case [25, 32]. Combining Eqs. 4 and 5 yields:

$$W_d = 3k_1\frac{\Omega}{b}\varepsilon_{IKB} = \frac{m_2}{m_1}\varepsilon_{IKB} \tag{6}$$

Since $3k_1\frac{\Omega}{b}$ can be experimentally determined, the estimation of Ω/b only requires knowledge of k_1 in Eq. 6. When the stress-strain curves are obtained and the plots of W_d vs. σ^2, ε_{NL} (non-linear strain; see Figure 1) vs. σ^2 and W_d vs. ε_{NL} are plotted, the second term in Eq. 6 can be used to estimate Ω/b, assuming $k_1 = 2$. On the other hand, m_1 and m_2 can be determined from the slopes of ε_{NL} vs. σ^2 or W_d vs. σ^2 plots, respectively (see below). Hence, if these assumptions are valid and if the micromechanism that is causing the dependence of ε_{NL} on σ (i.e. Eq. 4) is the *same* as the one responsible for W_d (Eq. 6), then the *ratio* m_2/m_1 should equal $3k_1\Omega/b$ as shown here and in Ref. [89].

Lastly, assuming the IKBs are cylinders with radii β_{av}, then the *reversible* dislocation density, ρ_{rev}, due to the IKBs is given by:

$$\rho_{rev} = \frac{2\pi N_k 2\alpha\beta_{av}}{D} = \frac{4\pi N_k \alpha\beta_{av}\gamma_c}{b} \tag{7}$$

where β_{av} is the average of β_{xc} and β_{yc}.

Experimental Details

The composites tested were made using two different techniques: HPing and MI. The former was used initially on the assumption – later proven incorrect – that Mg, like Al, would not spontaneously infiltrate a Ti_2AlC preform. Both techniques are described in detail below.

Hot pressing: The starting powders of Ti_2AlC (-325 mesh, 3-ONE-2, Voorhees, NJ) and Mg (-325 mesh, 99.8 % pure, Alfa Aesar, Ward Hill, MA) were ball-milled for 12 h and dried in a mechanical vacuum furnace at 150 °C for 24 h. The dried powder mixtures were poured and wrapped in graphite foil, that, in turn, were placed in a graphite die and HPed in a graphite-heated vacuum-atmosphere HP (Figure 4), (Series 3600, Centorr Vacuum Industries, Somerville, MA), heated at 10°C/min to 750°C and held at the target temperature for 1 h, after which the HP was turned off and the samples were furnace cooled. A load, corresponding to a stress of ~ 45 MPa, was applied when the temperature reached 500 °C and maintained thereafter. The samples were removed from the dies and the graphite foil was removed. These samples will be referred to as the "HP" samples.

Melt Infiltration: for the second set of samples, ≈ 50 vol.% porous preforms in the form of rectangular bars (1.2×1.2×70 cm^3) or cylinders (40 mm in diameter and 40 mm or 70 mm high) were fabricated by cold pressing the same Ti_2AlC powder together with ~ 1 wt. % polyvinyl alcohol as a binder at 45 MPa. Two microstructures were fabricated, random and oriented. The former were made by simply pouring, and cold pressing the Ti_2AlC-binder mixture into a steel die. To fabricate the latter, the Ti_2AlC-binder mixture was first poured into the die and manually vibrated for ~ 15 minutes in an attempt to orient the flaky Ti_2AlC powders perpendicular to the pressing direction [90].

The preforms' densities were calculated by dividing their weight by their volume because they were regularly shaped. For consistency, only those preforms that were 50±1% dense were used for the infiltration process (Figure 5).

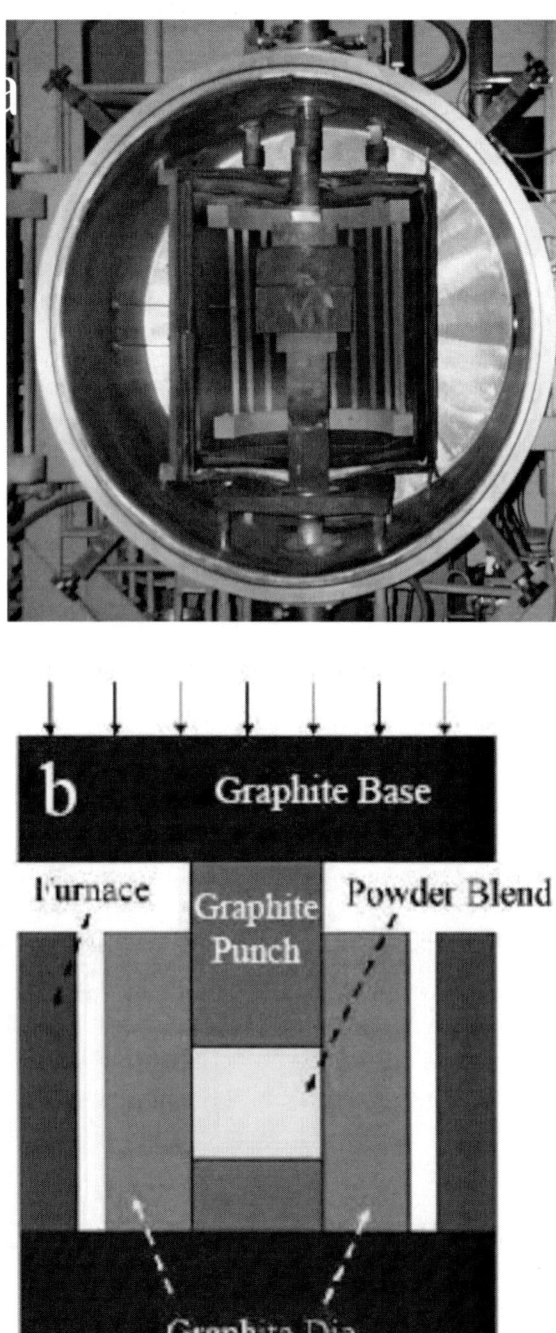

Figure 4. (a) Chamber of the graphite-heated vacuum-atmosphere HP, (b) schematic of the HPing system utilized to fabricate Mg-Ti2AlC composites using the Mg and Ti2AlC powders' blend.

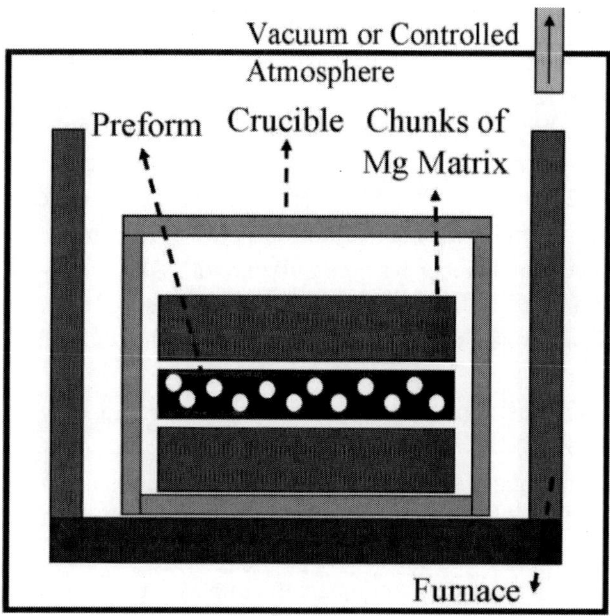

Figure 5. Schematic of the MI system utilized to fabricate Mg-Ti_2AlC composites.

The performs were then placed in a graphite-heated vacuum furnace and heated at 5°C/min to 900°C, held at the target temperature for 5 h, after which the furnace was turned off and the preforms were furnace cooled. More recent work showed that this heat treatment step is not necessary and can be eliminated.

To carry out the infiltration step, pure Mg chunks (99.8 % pure, Alfa Aesar, Ward Hill, MA) were used to surround the preforms that, in turn, were placed in alumina, Al_2O_3, crucibles (AdValue Technology, Tucson, AZ). The crucibles were covered with Al_2O_3 lids and placed in the same vacuum furnace used for sintering the preforms, heated at 10°C/min to 750°C, held at that temperature for 30 min, after which the furnace was turned off and the samples were furnace cooled. In all cases, the excess Mg surrounding the infiltrated preforms was machined off. These samples will henceforth be referred to as MI.

The composite samples were then annealed at 550° C for 6 h in flowing Ar in a tube furnace in order to investigate the thermal stability of the Mg matrices in the various composites.

Cylinders for compression tests *parallel* and *normal* to the cold-pressing direction were EDMed from the *same* oriented MI samples. Under compression, the basal planes in the former are *normal* to the loading direction, which is why these samples are referred to as "MI-N". When the basal planes are parallel to the loading direction, the samples will be referred to as "MI-P". This nomenclature is also valid for the Vickers hardness measurements because in the MI-N sample, the indenter is *normal* to the basal planes, etc. The randomly oriented samples will be referred to as MI-R. For clarity's sake, in most of the stress-strain figures, a small schematic of the relationship of the basal planes to the applied load is shown as an inset.

Also for the sake of comparison, Mg-50 vol.% Ti_3SiC_2 and Mg-50 vol. % SiC composites were fabricated by HPing. In this case, the starting powders were Ti_3SiC_2 (-325 mesh, 3-ONE-2, Voorhees, NJ), SiC (- 325 mesh, Alfa Aesar, Ward Hill, MA) and the same Mg powder used above. The processing details were identical to those of the Mg-Ti_2AlC (HP) composites described above. These samples will henceforth be referred to as "Mg-312" and "Mg-SiC", respectively.

Bulk Ti_2AlC and Ti_3SiC_2 samples were also made by hot isostatic pressing (HIPing). The starting powders were sealed in rubber bags under a mechanical vacuum and then cold isostatically pressed (CIPed) to ~ 250 MPa for ~ 5 min. The samples were then placed in crushed borosilicate glass and then in a hot isostatic press (HIP), heated to 750 °C at a rate of 5 °C/min, at which time the chamber was pressurized with Ar gas to ~ 100 MPa. The heating was then resumed at a rate of 10 °C/min to 1400 °C at which time the chamber was further pressurized to ~ 175 MPa and the samples were held for 2 h followed by furnace-cooling to room temperature.

The composite samples' microstructures were observed in a field emission scanning electron microscope, SEM, (Zeiss Supra 50VP, Germany) after cross-sectioning, mounting and polishing with a diamond solution down to 1 μm. The bulk Ti_3SiC_2 and Ti_2AlC samples were polished and etched for ~ 10 s with a 1:1:1 (volume) H_2O:HNO_3:HF etchant solution and their microstructures were then observed with an optical microscope, OM, (Olympus PMG-3, Tokyo, Japan). The oriented composite samples were cross-sectioned parallel and normal to the plate-like-grains in order to image the morphology in both directions (MI-P and MI-N).

X-ray diffraction (XRD) was carried out on bulk composites, powders and preforms in a diffractometer (Model 500D, Siemens, Karlsruhe, Germany) and the spectra were collected using step scans of 0.01° in the range of 10° to 90° 2 theta (2θ) and a step time of 2 s. Scans were made with Cu Kα radiation (40 KV and 30 mA).

The Vickers microhardness values, V_H, – measured using a microhardness indenter (LECO-M400, LECO Corp. St. Joseph, MI) – were determined by averaging at least 10 measurements at 1, 2, 3, 5 and 10 N. The hardness measurements were carried out on the MI (MI-R, MI-P and MI-N) and HP composites, pure polycrystalline Mg, dense Ti_2AlC, Ti_3SiC_2, Mg-SiC and Mg-312 composites.

The room temperature UCSs were measured using a hydraulic testing machine (MTS 810, Minneapolis, MN) (Figure 6) on small 4×4×4 mm^3 EDMed cubes. Six samples were tested. Tensile bars were EDMed from the random and oriented infiltrated preforms according to ASTM E8-04. Three samples were tested. EDMed cylinders 9.7 mm in diameter and 31 mm high were used to measure the Young's moduli in compression and to carry out the cyclic uniaxial compression tests. In all cases, the strains were measured by a capacitance extensometer (MTS, Minneapolis, MN) – attached to the samples (Figure 6) – with a range of 1 % strain. All the loading-unloading compression tests were performed in load-control mode at a loading-unloading rate of 15 MPa/s, respectively, which corresponds to a strain rate of ~ 2×10^{-4} s^{-1}.

The offset yield strength (Y_s) of the composites – being the stress required to produce a plastic deformation strain of 0.2 %, was determined by the stress corresponding to the intersection of the stress-strain curves offset by 0.2 %. These measurements were carried out at room temperature on small 4×4×4 mm^3 EDMed cubes, in displacement-control mode at a displacement rate of 0.005 mm/s.

TEM foils were prepared by a conventional TEM sample preparation process: 0.5 mm-thick slices were first cut from bulk samples using a low-speed diamond saw and further thinned with a disc-grinder to a thickness of about 20 μm. Final perforation was made by an ion mill operating at 5 kV. TEM characterization was performed using a field emission TEM (JEOL JEM-2010F) operating at 200kV. Images were collected with a multi-scan CCD digital camera. EDS analysis was carried out with an attached EDAX ultra-thin window X-ray energy dispersive spectrometer.

Also for the sake of comparison, Mg-50 vol.% Ti_3SiC_2 and Mg-50 vol. % SiC composites were fabricated by HPing. In this case, the starting powders were Ti_3SiC_2 (-325 mesh, 3-ONE-2, Voorhees, NJ), SiC (- 325 mesh, Alfa Aesar, Ward Hill, MA) and the same Mg powder used above. The processing details were identical to those of the Mg-Ti_2AlC (HP) composites described above. These samples will henceforth be referred to as "Mg-312" and "Mg-SiC", respectively.

Bulk Ti_2AlC and Ti_3SiC_2 samples were also made by hot isostatic pressing (HIPing) for the sake of comparison. The starting powders were sealed in rubber bags under a mechanical vacuum and were cold isostatically pressed (CIPed) to ~ 250 MPa for ~ 5 min. The samples were then placed in crushed borosilicate glass and then in a hot isostatic press (HIP), heated to 750 °C at a rate of 5 °C/min, at which time the chamber was pressurized with Ar gas to ~ 100 MPa. The heating was then resumed at a rate of 10 °C/min to 1400 °C at which time the chamber was further pressurized to ~ 175 MPa and the samples were held for 2 h followed by furnace-cooling to room temperature.

Figure 6. Front view of the hydraulic testing machine used for cyclic compression tests; also shown are the sapphire extension rods attached to the surface of the sample to measure its strain during loading.

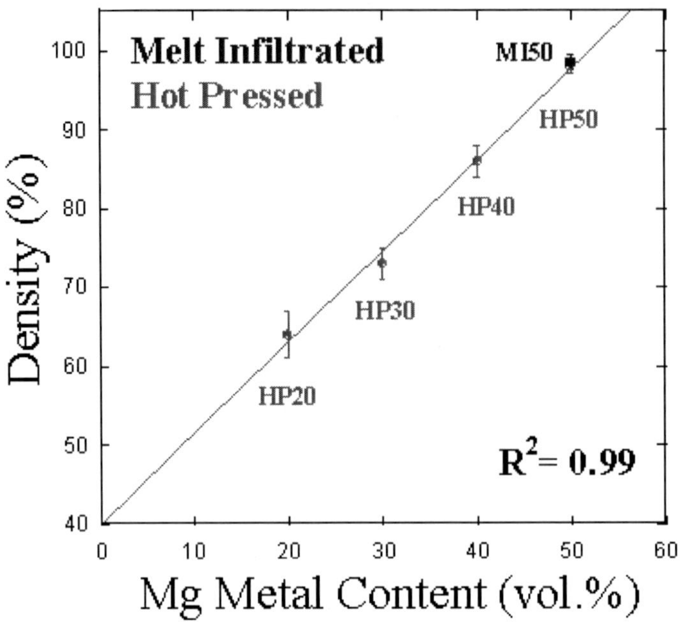

Figure 7. Evolution of density versus Mg metal content in HPed and melt-infiltrated composites.

Results and Discussion

1) Density Measurements

The highest density of 2.87±0.03 Mg/m^3 (~ 98.5 % of theoretical) in the HP composites was only obtained when the Mg content was 50 vol. % (HP50). At 40 vol. % Mg the density was ~ 85 % of theoretical (HP40). Lower Mg contents resulted in more porous samples that were not studied further (Figure 7). The theoretical density was calculated assuming the densities of Mg and Ti_2AlC to be 1.74 Mg/m^3 and 4.11 Mg/m^3 respectively [16]. At 2.87±0.05 Mg/m^3, the densities of the Mg-50 vol.% Ti_2AlC MI samples were also ~ 98.5 % of theoretical.

2) Microstructural Results

The microstructure of the HP50 and MI50 samples was quite homogeneous (Figure 8 a and b). As expected, the microstructure of the MI50-oriented sample, however, was different in the two different directions (Figure 8 c and d) being MI-P and MI-N, respectively. It is readily observed that manual vibration of the powder prior to cold pressing oriented most of the plate-like-grains with their basal planes exposed to the surface in MI-P sample. In contradistinction, most of these grains' basal planes are perpendicular to the surface in MI-N sample. The width and the thickness of the average Ti_2AlC grains were, respectively, 30±10 and 5±3μm.

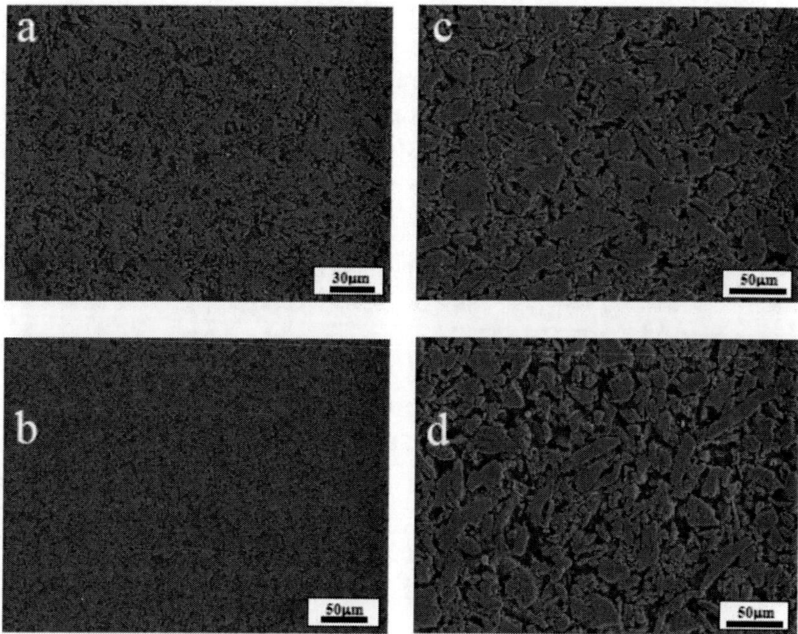

Figure 8. Secondary electron SEM image of polished surface of, a) HP50 and, b) MI-R, c) MI-P and, d) MI-N composites.

Figures 9 a and b show the fractured surfaces of HP50 and MI50 composites, respectively. Very few facetted Mg single crystals were formed throughout the entire microstructure of both MI50 and HP50 composites. Interestingly enough, similar, but much larger, facetted single crystals were formed on the surface of the Al_2O_3 lids during MI (Figures 9 c and d) used to keep the Mg from evaporating. It follows that they were formed by an evaporation/condensation process. Further work showed that the morphology of the Mg crystals formed on various substrates is significantly different and is a strong function of the type of substrate used.

Figure 10 shows several TEM images of a HP50 composite sample, where the Mg matrix appeared to have completely wet the Ti_2AlC. In some areas some porosity was observed. Presence of areas where the Mg matrix appears to have formed as particles, most likely similar to those single crystals that were shown on the fractured surfaces is also evident.

Figures 10 c to f show TEM images taken at higher magnifications from regions similar to those shown in Figure 8. They all show areas where Mg matrix appears to have wet the Ti_2AlC matrix and areas where Mg single crystals were formed in the open spaces and porosities of the microstructure. Note that these finite porosities were only observed in TEM images. It is thus reasonable to assume that the single crystals observed on the fractured surfaces are similar to those observed in the TEM images.

Figures 11 a to d show TEM images of the MI50 composite. Similar to the HP50 composite, the molten matrix wet and fully infiltrated the preform. There is, however, less porosity in this microstructure and thus limited regions filled with Mg single crystals. Again, this is in good agreement with the fractured surface meaning that the Mg single crystals observed on the fractured surfaces are similar to those in the TEM images.

Figure 12 a shows the formation of kink bands (the governing phenomenon in the deformation of KNE solids) in a single grain of Ti_2AlC on the polished surface of a HP50 composite. Figures 12 b and c depict the formation of kink bands with very sharp radii of curvature on a fractured surface of a MI50 composite. Based on these micrographs, and the evidence shown below, it is reasonable to assume that these kink bands were preceded by IKBs.

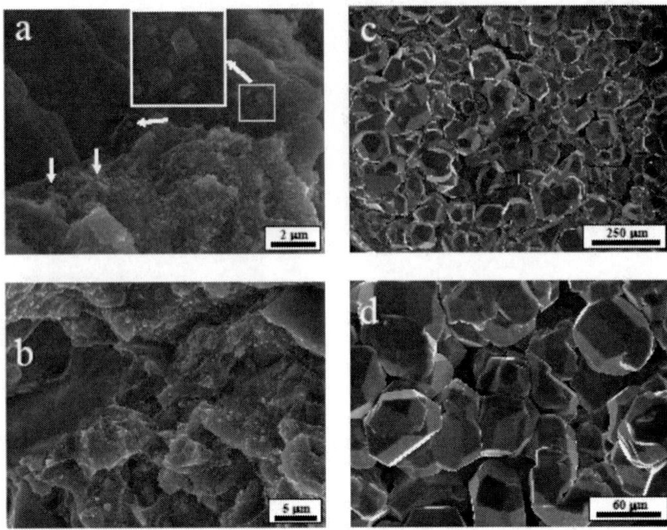

Figure 9. SEM image of fractured surface of, a) HP50; inset shows the facetted Mg single crystals at higher magnifications, and b) MI50 composites; c) and d) Facetted Mg single crystals formed during MI on the surface of the alumina lids.

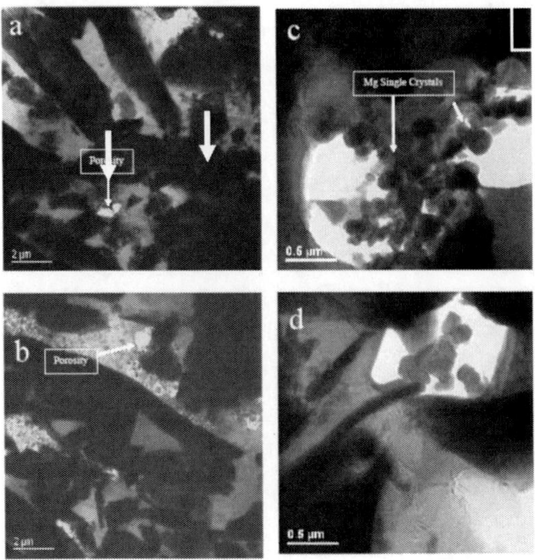

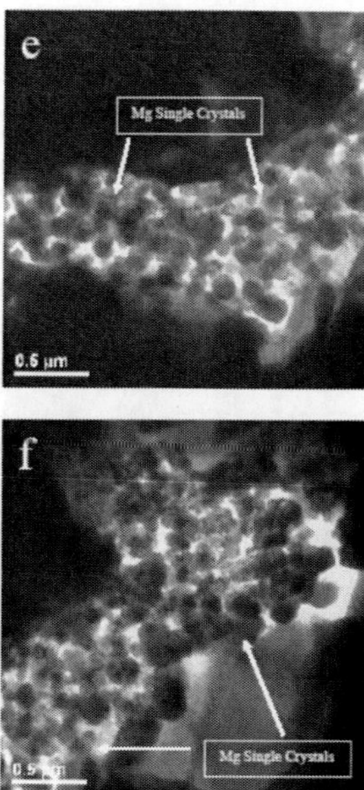

Figure 10. TEM images of HP50 composite showing the presence of areas where Mg matrix appears to have wet the Ti_2AlC matrix, very small amounts of porosity and areas where Mg single crystals were formed in the open spaces and porosities of the microstructure.

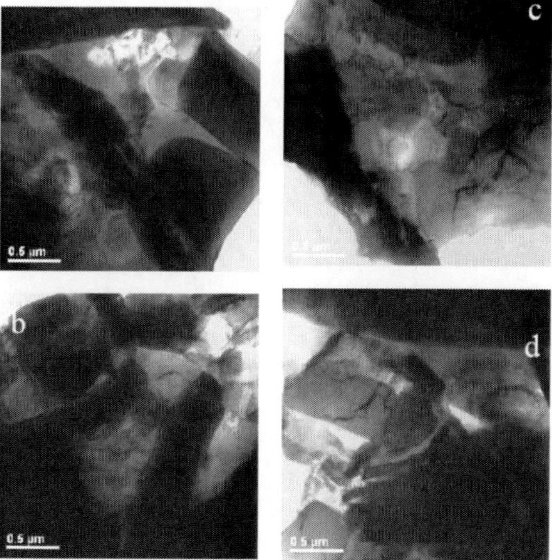

Figure 11. TEM images of MI50 composite wherein the molten matrix has wet and fully infiltrated the preform. Compared to HP50 composite less porosity and limited regions filled with Mg single crystals are observed.

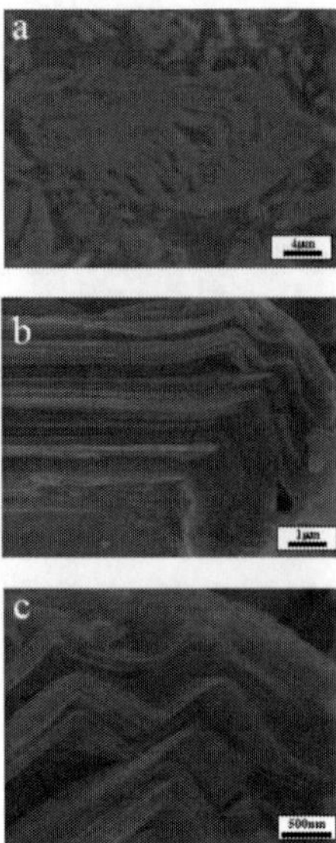

Figure 12. Formation of kink bands with very sharp radii of curvature in, a) a single grain of Ti_2AlC on the polished surface of the HP50 composite sample and, b and, c) fractured surfaces of MI50 composite sample.

3) X-ray Diffraction Results

Typical XRD patterns of the HP50 and MI50 composites (Figure 13 a) contained peaks for Ti_2AlC, Mg, TiC (~ 5 vol. % impurity in the starting Ti_2AlC powder), MgO and Si (the latter added as an internal standard). Figure 13 b also shows the XRD pattern of HP50 and MI50 samples before and after annealing, for the sake of comparison, containing peaks for Ti_2AlC, Mg, TiC (~ 5 vol. % impurity in the starting Ti_2AlC powder) and MgO.

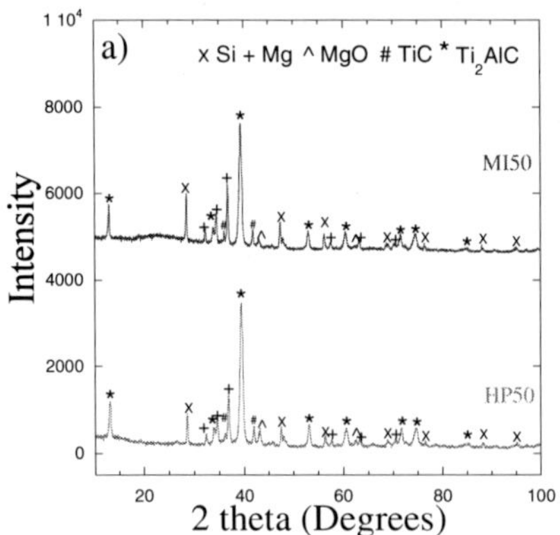

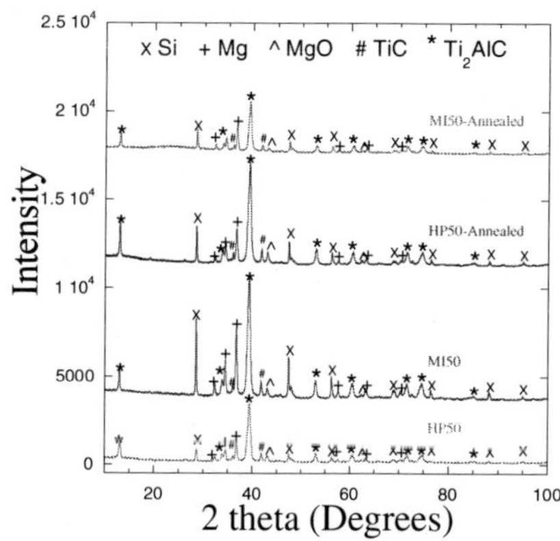

Figure 13. a) X-ray diffraction pattern of Mg-50 vol.% Ti2AlC composite fabricated by HPing and MI; b) X-ray diffraction pattern of Mg-50 vol.% Ti2AlC composite fabricated by HPing and MI before and after annealing (carried out at 550° C for 6 h in flowing Ar of a tube furnace), containing peaks for Ti2AlC, Mg, TiC (~ 5 vol. % impurity in the starting Ti2AlC powder) and MgO. In all cases Si was added as an internal standard.

The ratio of XRD peak intensities of the (002) basal planes to (103) planes in Ti_2AlC and the (002) basal planes to (104) planes in Ti_3SiC_2, respectively, were obtained from the XRD spectra of their corresponding starting powders, XRD cards [91, 92] and, oriented preforms. The data is summarized in Table 2. A perusal of this table makes it clear that the ratios of the 002/101 intensities in the oriented Ti_2AlC and Ti_3SiC_2 preforms are several multiples of what they are in the as-received powder or according to the XRD cards [91, 92]. We note in passing that one reason for the ease by which these powders can be aligned is their flake-like nature (Figure 14 a) [90]. Also note the equiaxed morphology of the Ti_3SiC_2 grains (Figure 14 b). It is explained below how the difference in grain morphology of Ti_2AlC and Ti_3SiC_2 grains, best manifested by comparing the OM micrographs of Figure 14, will render the latter relatively less amenable to kinking as compared to the plate-like grains of the former.

Table 2. The ratio of XRD peak intensities of (002) basal planes to (103) planes in Ti_2AlC and (002) basal planes to (104) planes in Ti_3SiC_2, respectively, obtained from their corresponding starting powders, XRD cards and those of the oriented preforms; in the latter case, XRD was performed on the surface of the preforms, wherein the majority of the basal planes were normal to the cold-pressing direction during preform preparation. Note that this direction, as described herein for the composites materials, is analogous to the MI-P composite sample

↓Intensities– Material→	Ti_2AlC powder	Ti_2AlC (XRD card [91])	Oriented Ti_2AlC preform
I (002) / I (103)	0.5	0.4	1.3
↓Intensities– Material→	Ti_3SiC_2 powder	Ti_3SiC_2 (XRD card [92])	Oriented Ti_3SiC_2 preform
I (002) / I (104)	0.4	0.2	1.7

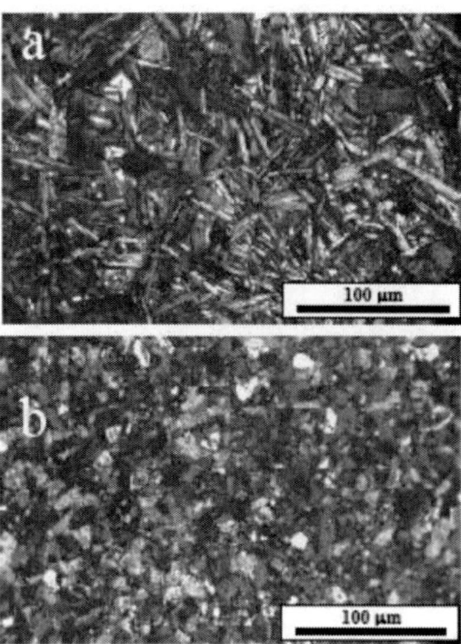

Figure 14. Optical micrographs of fully dense, polished and etched, a) Ti_2AlC, and b) Ti_3SiC_2. Both samples were fabricated by cold isostatically pressing (CIPing) to ~ 250 MPa for ~ 5 min followed by hot isostatic pressing (HIPing) at 1400 ºC and ~ 175 MPa for 2 h followed by furnace-cooling to room temperature.

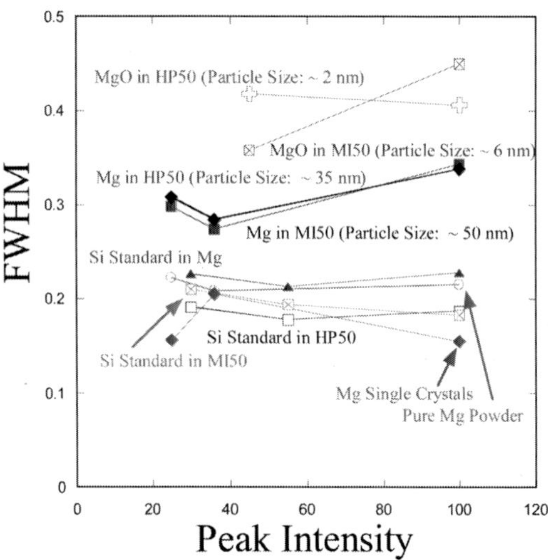

Figure 15. FWHM of Mg and MgO vs. peak intensity; three of the high intensity peaks were used in each case and compared with those of Si standard, pure Mg powder and Mg single crystal peaks.

When the full-widths at half maximum, FWHM, of the Mg peaks in the HP50 and MI50 composites are compared with those of the pure Mg powder ($d_{av} \approx 150$ µm), Mg single crystals or Si (Figure 15) it is apparent that the former are significantly broader. Using the Scherrer formula [93] the Mg particle size was estimated to be ~35±15*nm* in both HP50 and MI50 composites.

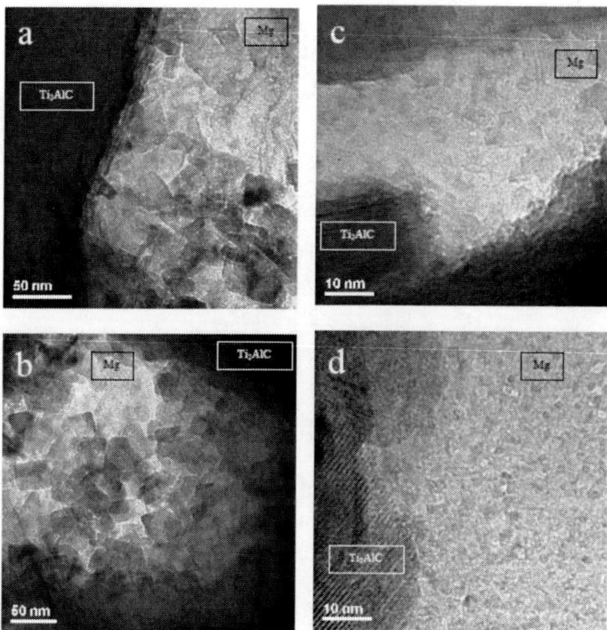

Figure 16. TEM images of, (a) and (b), the MI50 composite and, (c) and (d), the HP50 composite showing the presence and morphology of the nc-Mg matrices within the composites.

More TEM images from the Mg matrix of the MI50 (Figure 16 a and b) and HP50 (Figure 16 c and d) composites were taken at higher magnifications in order to confirm the presence of nano-crystalline Mg matrix. From these micrographs it is evident that the molten Mg matrix solidified in the form of nano-crystals (≈20 nm in MI50 and ≈10 nm in HP50).

Survival of these Mg nanocrystals after 1 h at 750°C in the HP vacuum implies the presence of a potent grain-growth inhibitor. The presence of MgO peaks in the XRD spectra strongly suggests that MgO plays that role. Based on the FWHM of the MgO peaks its grain size is estimated to be of the order of ~ 3±1 *nm* in both the HP50 and MI50 composites.

These microstructures were also *remarkably stable*: annealing at 550 °C for 6 h did *not* result in grain growth as evidenced by the FWHM of the Mg-peaks after annealing (Figure 17). This remarkable result – most probably associated with the presence of MgO at the grain boundaries as a potent grain-growth inhibitor – bodes well for possible applications of this composite at elevated temperatures.

Energy dispersive X-Ray Spectroscopy (EDS) in the TEM of the Mg matrix in several regions similar to those shown in Figure 16 confirmed the presence of Mg, Ti and O. EDS microanalysis of the Mg matrix in the SEM revealed the presence of ~ 3±1 at. % Ti. It is likely that Ti diffuses out of the Ti_2AlC grains into the Mg matrix and Mg diffuses in.

It is important to note, however, that according to the Mg-Ti binary phase diagram [94] the solubility of Ti in Mg at 750°C is almost nil. The absence of pure Ti regions in the TEM, however, suggests that the Ti is supersaturated in the Mg matrix. Further work [95] suggested that Ti may be also present in the form of rutile and/or anatase.

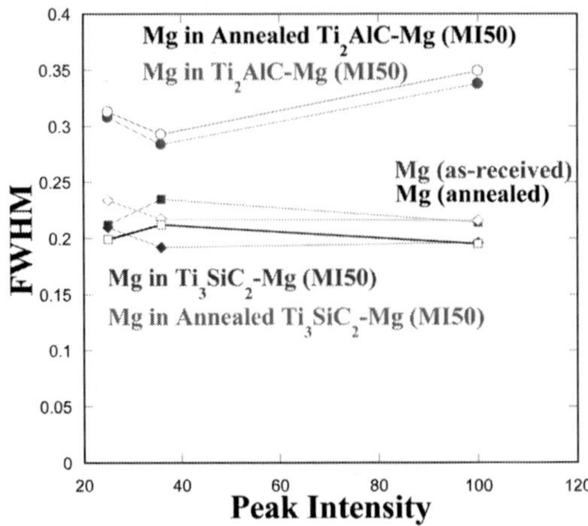

Figure 17. FWHM of Mg peaks in Mg-Ti_2AlC (MI50) composite before and after annealing; included in the same figure, for the sake of comparison, is the FWHM of the Mg ingot and the Mg in Mg-Ti_3SiC_2 (Mg-312) composite before and after annealing; the three highest intensity peaks were used in each case.

Interestingly and consistent with what is known about how strongly bound the C is in the MAX phases [96] the C does not appear to diffuse out of the Ti_2AlC grains. This conclusion is also in agreement with the fact that the MAX phases do not melt congruently, but dissociate into $M_{n+1}X_n$ and an A-rich liquid [9, 97]. It is also consistent with how Ti_3SiC_2 reacts with other reactive liquids. For example, when Ti_3SiC_2 is immersed in molten Al or crolyite, the Si diffuses out leaving behind a TiC-rich phase [98, 99]. In other words, in neither case does the C-diffuse out.

EDS microanalysis of the Ti_2AlC grains in the TEM and SEM revealed the presence of Mg within them (Figure 18 a). The sum of Mg and Ti concentrations at various distances from the grain edges was ~ 50 at. %; it follows that the solubility of Mg in Ti_2AlC is non-negligible. In other words, the solid solution $(Ti_{1-x}Mg_x)_2AlC$, in which x is at least as high as 0.2, might exist.

Due to the fairly low Mg-content, its effect on the c-lattice parameter of Ti_2AlC grains is small and, within experimental scatter, identical to the as-received powders (Figure 18 b). The increase in a-lattice parameter, on the other hand can be attributed with the larger radius of Mg in comparison with Ti. Note the values reported herein are, most probably, *not* equilibrium values. The diffusion coefficient of Mg in Ti_2AlC at 750 °C is estimated (x^2/Dt, where x is the distance from Mg/Ti_2AlC interface and t is the diffusion time) to be $\approx 3\times10^{-16}$ m^2/s.

In order to further investigate the variation of lattice parameters in several MAX phases when in contact with molten Mg, and to study the stability of MAX phases in the presence of Mg melt, we attempted to fabricate Mg-Ti_2AlC, Mg-Nb_2AlC and Mg-Ti_3SiC_2 composites by MI, similar to the manner explained in the experimental section above.

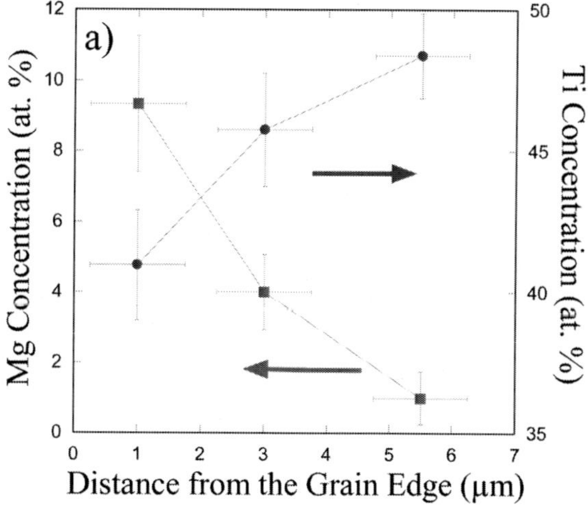

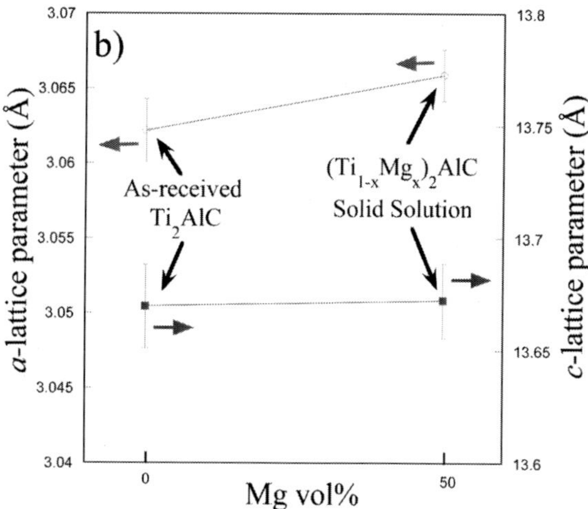

Figure 18. a) Concentration in at.% of Mg and Ti within the Ti2AlC grains verifying the formation of a (Ti1-xMgx)2AlC solid solution, with an x as high as 0.2; b) Variation of a and c lattice parameters in as-received Ti2AlC and that within the MI composite.

In order to obtain equilibrium values, we fabricated the composites for two different soaking times: for the first set of samples, we soaked the porous MAX phase preforms for 1 h in the Mg melt at 750 ºC, followed by furnace cooling. The second set of samples were fabricated by soaking the preforms for 4 h at 750 ºC, also followed by furnace cooling. For each of the samples, the *a* and *c* lattice parameters of the MAX phase were calculated and plotted as a function of soaking time.

Figure 19 a shows the *a* and *c* lattice parameters of Ti_2AlC. Within the experimental scatter, the *a* lattice parameter – as mentioned earlier – seem to be increasing with soaking time up to 1 hr and then apparently reaches an equilibrium value at 4 hrs. The *c* lattice parameter seems to remain unchanged and equal to its value in the starting powders.

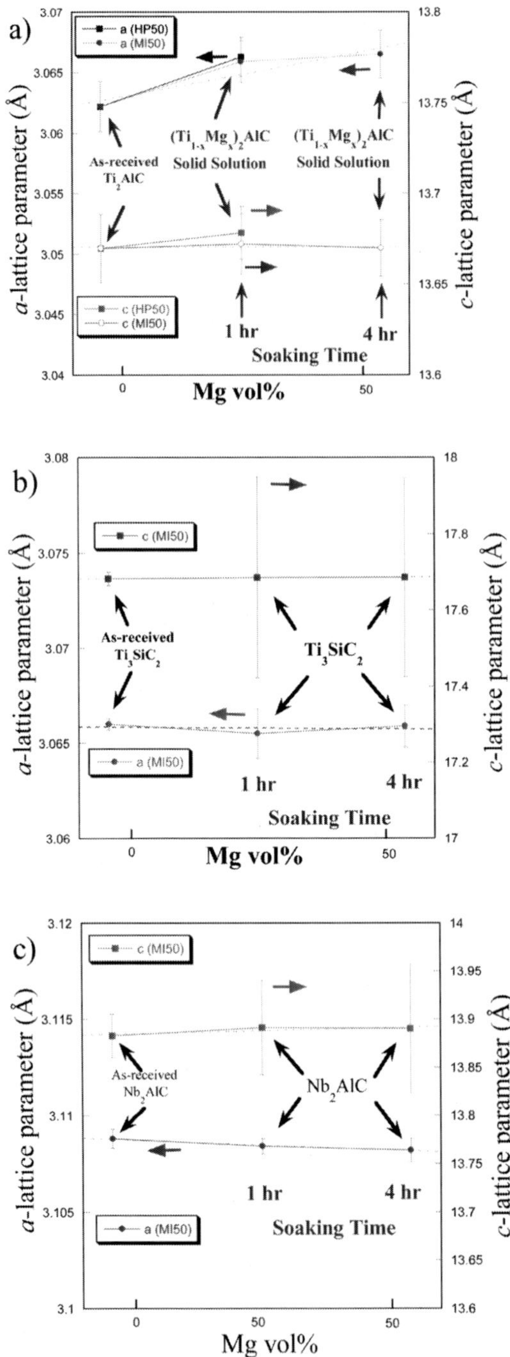

Figure 19. Variation of *a* and *c* lattice parameters of as-received, (a) Ti_2AlC, (b) Ti_3SiC_2, and (c) Nb_2AlC and those within their MI composites after soaking for 1 hr and 4 hrs in the Mg melt.

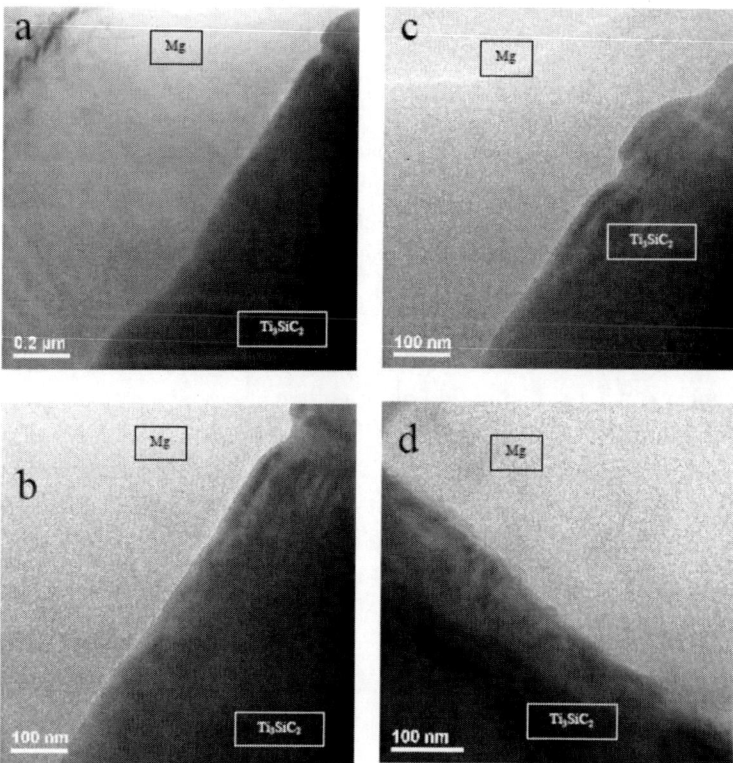

Figure 20. TEM images of Mg-50 vol.% Ti_3SiC_2 composite fabricated by MI at 750° C for 1 h, showing the absence of nanocrystalline Mg matrix within the composite.

On the other hand, Figure 19 b and c show the *a* and *c* lattice parameters of Ti_3SiC_2 and Nb_2AlC, respectively. Apparently, within the experimental scatter, both *a* and *c* lattice parameters of Ti_3SiC_2 and Nb_2AlC, within their composites, seem to be constant as the soaking time increases.

These results show the stability of Ti_3SiC_2 and Nb_2AlC MAX phases when in contact with Mg melt, which is probably why the Mg matrix did not appear to be in nanoscale in Mg-Ti_3SiC_2 and Mg-Nb_2AlC composites. Figure 20 shows the TEM images obtained from the Mg-Ti_3SiC_2 composites, showing the absence of a nanocrystalline Mg matrix, corroborating the results shown in Figure 19.

4) Effect of Texture on the Mechanical and Damping Properties of "MAXMET"s

a) Results

Not surprisingly, and similar to all other MAX phases [4, 25, 100], and Mg, both HP and MI composites are readily machinable even with a manual hack-saw with no lubrication or cooling. They can also readily be EDMed with a significantly higher rate and ease than the MAX phases. According to the machinist, their machinability is similar to 7000 series Al alloys.

The effect of indentation loads on the V_H values of the HP, MI-R, MI-P and MI-N samples, together with those of fully dense Ti_2AlC, pure Mg, Mg-312 and Mg-SiC for the sake of comparison are plotted in Figure 21; Figure 22 shows secondary electron SEM images of Vickers indentation marks in the MI and HP composites.

Figure 23 a compares the UCS of all materials tested in this work. Figures 23 b and c show the compressive stress-strain curves to the point of fracture indicating offset yield strength and fracture point in HP50 and MI50 composites, respectively.

For cyclic compression tests, typically five cycles are obtained at each load. For the most part, the first cycles were very slightly open, registering a plastic strain of the order of ~ 0.05%.

However, all subsequent cycles, to the same stress, were closed and exceptionally reproducible, which is why in Figure 24 only one loop at any given stress is plotted. Typical stress-strain loops at various stresses for the MI-P (Figure 24 a), MI-N (Figure 24 b) and MI-R (Figure 24 c) – loaded to roughly ~ 75% of their UCS – at different stresses are all closed. Typical fully reversible stress-strain loops for the Mg-312 and Mg-SiC composites (Figure 24 d) are, however, significantly smaller than the rest. Figures 24 e and 24 f show typical fully reversible stress-strain loops of HP50 and HP40 composites, respectively.

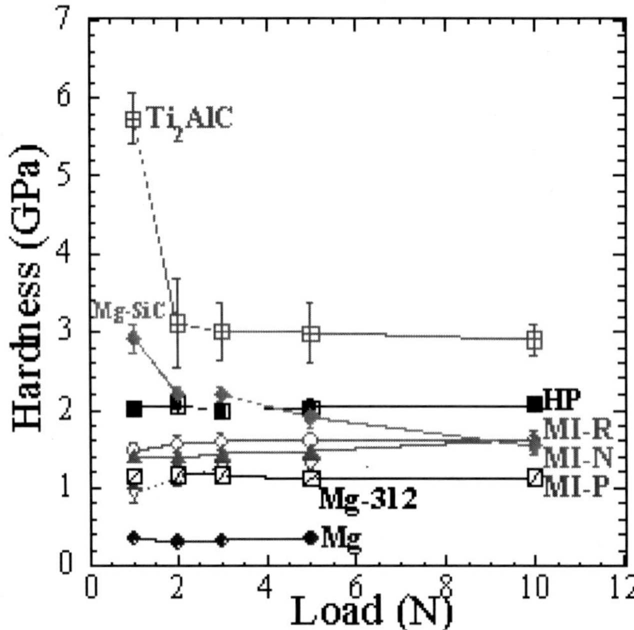

Figure 21. Effect of indentation loads on the V_H values of the HP, MI-R, MI-P and MI-N samples, together with those of fully dense Ti_2AlC, pure Mg, Mg-312 and Mg-SiC for comparison.

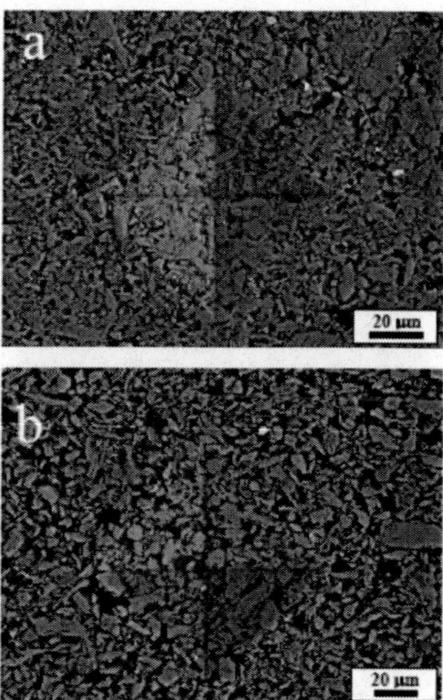

Figure 22. SEM micrographs of Vickers indentations at 10 N on a) HPed and b) MI Mg-Ti_2AlC composites.

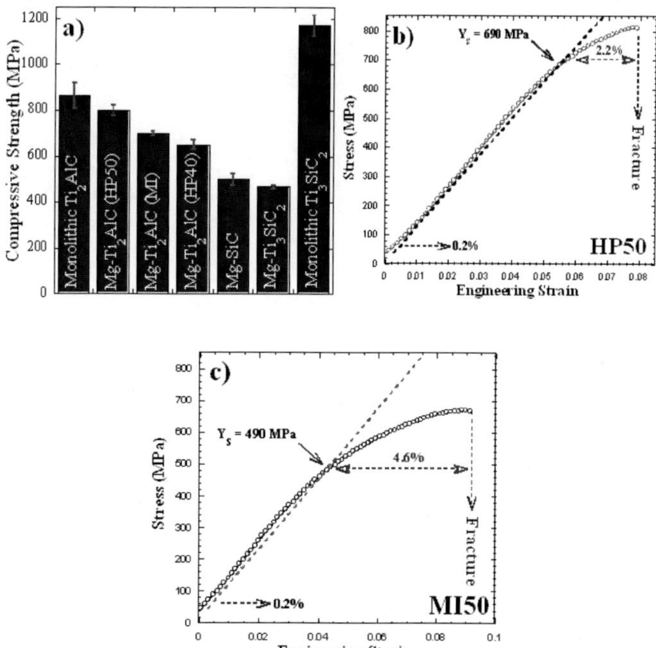

Figure 23. (a) Plot of UCS values of all materials tested in this work; compressive stress-strain curves indicating offset yield strength and fracture point for, (b) HP50 and, (c) MI50 composites.

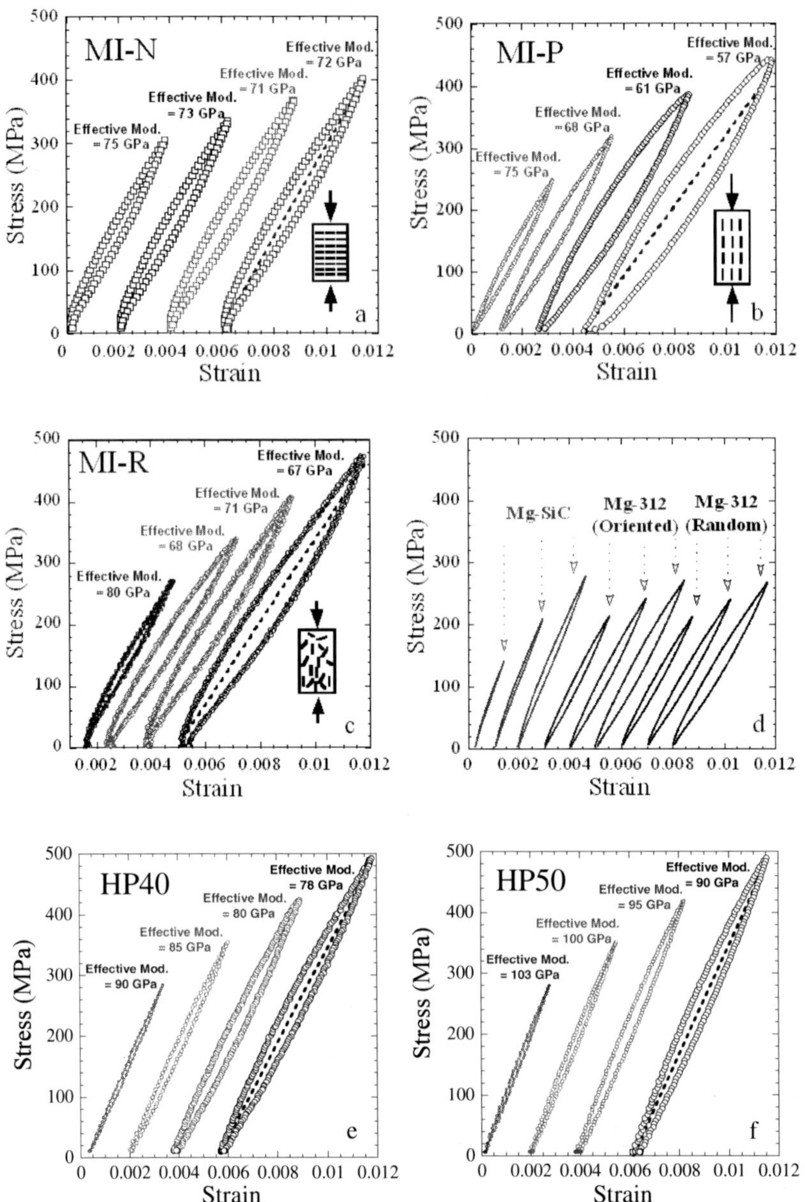

Figure 24. Typical compressive stress-strain curves for a) MI-N, b) MI-P, c) MI-R, d) Mg-SiC and Mg-312, e) HP40 and, f) HP50 composites. Only the fifth cycle is shown and the curves are shifted horizontally for clarity.

Defining a Young's modulus for loops such as those shown in Figure 24 is problematic. However, to obtain an *approximate* "effective" Young's modulus, $\overline{E}$, least squares fits of the entire data set that resulted in diagonal lines bisecting the loops (only those at the highest loads are shown in figures) were carried out at each stress. The results are summarized in Table 3.

Table 3. Effective Young's modulus, $\overline{E}$, for MI-P, MI-R, MI-N, HP50, HP40, Mg-312, Mg-SiC, Ti_2AlC and Ti_3SiC_2 samples tested. Also listed are the values for the *upper* and *lower* bounds of E [E(u) and E(l)], G [G(u) and G(l)] and Poisson's ratio, ν, [ν(u) and ν(l)], respectively

Material	MI-P	MI-R	MI-N	HP50	HP40
$\overline{E}$ (GPa)	69±10	72±6	74±3	88±5	83±5
E(u)	130±3	130±3	130±3	130±3	127±3
E(l)	74±1	74±1	74±1	74±1	-
G(u)	69	69	69	69	67
G(l)	33	33	33	33	-
ν(u)	0.27	0.27	0.27	0.27	0.24
ν(l)	0.25	0.25	0.25	0.25	-

Material	Mg-312	Mg-SiC	Ti_2AlC	Ti_3SiC_2
$\overline{E}$ (GPa)	74±4	117±17	218±6	237±22
E(u)	140±10	260	-	-
E(l)	75±1	82	-	-
G(u)	82	106	-	-
G(l)	34	35	-	-
ν(u)	0.27	0.25	-	-
ν(l)	0.25	0.20	-	-

Based on the KNE model, the mechanical hysteresis of a KNE solid can be characterized by three parameters, σ, ε_{NL} and W_d – all obtainable from the hysteretic stress-strain curves shown in Figure 24 – listed in Table 4 for the samples tested.

According to the KNE model [34-36, 43, 48], plots of W_d vs. σ^2, ε_{NL} vs. σ^2 and W_d vs. ε_{NL} should all yield straight lines. With the notable exception of the Mg-SiC composite (see below), that is what was observed (Figures 25 a-f). The lowest correlation coefficient, R^2, obtained from least squares analysis of the results, with again the exception of the SiC-containing system, was > 0.95.

Table 5 lists the threshold stresses, σ_t, obtained from the W_d vs. σ^2 plots, viz. Figure 25 a and d. Also, based on the results shown in Figures 25a-f, the KNE model, and the constants listed in Table 5, the values of 2α, Ω/b, N_k, $2\beta_{av,c}$, $2\beta_{av}$, ρ_{rev} and ε_{IKB} were calculated. It is important to note that the values of 2α in Table 5 are calculated from the σ_t values (Table 5) and Eq. 1, assuming M=3. The latter is a good first assumption. The ε_{NL} values labeled "*measured*" are those measured by the extensometer directly attached to the samples' surface.

Table 4. List of measured stress (σ), nonlinear strain (ε_{NL}), and dissipated energy (W_d) for MI-P, MI-R, MI-N, HP50, HP40, Mg-SiC, Mg-312 and randomly oriented Ti_2AlC and Ti_3SiC_2 samples tested. Also listed are the m_1, m_2 and their ratio and $3k_1\Omega/b$ values obtained from the slopes of ε_{NL} vs. σ^2, W_d vs. σ^2 and W_d vs. ε_{NL} plots, respectively

	σ MPa	ε_{NL}	W_d MJ/m^3	m_1 (MPa)$^{-2}$	m_2 (MPa)$^{-1}$	m_2/m_1 (MPa)	$3k_1\Omega/b$ (MPa)
MI-P	250	0.0007	0.0832	1.5×10^{-8}	3.4×10^{-6}	230	229
	319	0.0012	0.1922				
	388	0.0021	0.3443				
	445	0.0027	0.5850				
MI-R	275	0.0007	0.0795	9.8×10^{-9}	2.2×10^{-6}	225	223
	340	0.0011	0.1755				
	410	0.0017	0.2752				
	475	0.0021	0.4171				
MI-N	305	0.0005	0.1101	9.0×10^{-9}	2.0×10^{-6}	225	224
	338	0.0007	0.1440				
	370	0.0009	0.1962				
	405	0.0011	0.2573				
HP50	280	0.0001	0.0324	7.7×10^{-9}	1.8×10^{-6}	235	237
	350	0.0003	0.0705				
	420	0.0006	0.1352				
	490	0.0013	0.2862				
HP40	285	0.0002	0.0398	6.4×10^{-9}	1.5×10^{-6}	231	232
	355	0.0004	0.0848				
	423	0.0008	0.1498				
	492	0.0012	0.2775				
Ti_2AlC	342	0.0002	0.0166	1.0×10^{-9}	2.4×10^{-7}	230	229
	445	0.0003	0.0359				
	537	0.0004	0.0557				
	628	0.0005	0.0794				
Mg-SiC	70	0.0001	0.0005	1.3×10^{-8}	4.2×10^{-7}	32	32
	140	0.0004	0.0045				
	210	0.0006	0.0168				
	280	0.0009	0.0459				
Mg-312 (Random)	159	0.0001	0.0178	1.8×10^{-8}	1.7×10^{-6}	94	93
	186	0.0002	0.0304				
	213	0.0003	0.0457				
	241	0.0005	0.0671				

Mg-312 (Oriented)	159	0.0003	0.0179	1.7×10^{-8}	1.7×10^{-6}	102	99
	188	0.0004	0.0274				
	215	0.0006	0.0432				
	243	0.0008	0.0660				
Ti_3SiC_2	307	0.0001	0.0208	2.8×10^{-9}	5.2×10^{-7}	193	192
	417	0.0003	0.0516				
	514	0.0006	0.0988				
	623	0.0009	0.1732				

b) Discussion

KINKING NONLINEAR ELASTICITY

The most important result of this work is the exceptional damping capability of the MI-P and HP50 composites. The W_d's of the MI-P and HP50 composites are ~0.6 MJ/m^3, a value that surpasses the previous record of 0.42 MJ/m^3 at 450 MPa reported in [44] for MI-R composite, by almost 50 % [37]. Note that the MI-P value was achieved at 450 MPa, whereas the HP50 value was achieved at 610 MPa.

When the W_d results of MI-P composite are compared with those of fully dense single-phase Ti_2AlC with comparable grain size, it is evident that the former are higher by at least one order of magnitude. Also when the W_d results of MI-P composite are compared with those of fully dense and 10 vol.% porous Ti_2AlC [26] with significantly larger grains (d_{av}=113±60μm and 2α=14±7μm for the dense sample and d_{av}=133±70 μm and 2α=16±7μm for the porous sample [26]), at ~ 350 MPa, W_d of the composites fabricated are larger by at least a factor of ~ 2 and 4, respectively. Note that W_d is a strong function of grain size and that the grains in the previous work [26] were considerably larger than the ones explored. Furthermore, as argued elsewhere [101], and confirmed, the relationship between Ω/b and grain size is essentially Hall-Petch like (see below).

The response of the Mg-Ti_2AlC composites strongly depended on the orientation of the basal planes relative to the loading direction (Figure 24) in a way that is consistent with a kinking phenomenon. Recall that kinking is a plastic instability and should be greatly enhanced if the basal planes are loaded edge-on, compared to when they are loaded along the c-axis. It is this simple intuitive conjecture that explains why the W_d's of MI-P samples are roughly double those of the MI-N samples, with those associated with MI-R in between (Figure 25).

Table 5. List of experimentally measured σ_t values obtained from the W_d vs. σ^2 plots (Figure 25 a and d) and 2α values calculated using the σ_t values in column 1 and Eq. 1 assuming M=3. Also listed are calculated values of Ω/b obtained from Eqs. 4 & 5 and 6 (columns 4 and 5, respectively), N_k, $2\beta_{av,c}$, ε_{IKB} calculated from the third term of Eq. 4 and ε_{NL} measured directly by the extensometer. The $2\beta_{av}$ and ρ_{rev} values at the stress levels listed in the last column are also included. For all cases, b = 3.0 Å, M = 3, w = 5b, k_1 = 2; G and ν of the Mg-Ti_2AlC, Mg-Ti_3SiC_2 and Mg-SiC composites were assumed to be ~ 51 GPa and 0.26, ~ 58 GPa and 0.26, and ~ 70 GPa and 0.22 (Table 3). Those of Ti_2AlC [26], Ti_3SiC_2 [101], SiC [102], and Mg [35] are *118* GPa and *0.2*, *144* GPa and *0.2*, *192* GPa and *0.142*, and *19* GPa and *0.35*, respectively

	σ_t (MPa)	2α (μm)	Ω/b (MPa) Eqs. 4&5	Ω/b (MPa) Eq. 6	N_k (m^{-3})	$2\beta_{av,c}$ μm)	$2\beta_{av}$ μm)	ρ_{rev} (m^{-2})	ε_{IKB} calculated	ε_{NL} measured	σ (MPa)
MI-P	162	3	38.4	38.2	4.7×10^{17}	0.29	0.79	1.5×10^{14}	0.0025	0.0027	445
MI-R	198	2	37.5	37.2	1.1×10^{18}	0.23	0.56	1.6×10^{14}	0.0018	0.0021	475
MI-N	198	2	37.5	37.3	9.7×10^{17}	0.23	0.48	1.3×10^{14}	0.0011	0.0011	405
HP50	225	2	39.2	39.5	1.8×10^{18}	0.20	0.45	1.7×10^{14}	0.0015	0.0013	490
HP40	219	2	38.5	38.7	1.3×10^{18}	0.21	0.48	1.3×10^{14}	0.0012	0.0012	492
Ti_2AlC (Random)	226	8	38.3	38.2	7.1×10^{15}	0.48	0.95	8.8×10^{12}	0.0003	0.0002	445
Mg-SiC	99	18	5.6	5.4	5.2×10^{15}	0.65	1.85	2.2×10^{13}	0.0009	0.0009	280
Mg-312 Random	128	7	15.7	15.5	8.7×10^{16}	0.41	0.77	5.5×10^{13}	0.0007	0.0005	241
Mg-312 Oriented	134	7	16.9	16.6	1.0×10^{17}	0.40	0.71	5.7×10^{13}	0.0007	0.0008	243
Ti_3SiC_2 (Random)	284	10	32.1	31.9	3.4×10^{16}	0.47	0.68	2.8×10^{13}	0.0003	0.0003	417

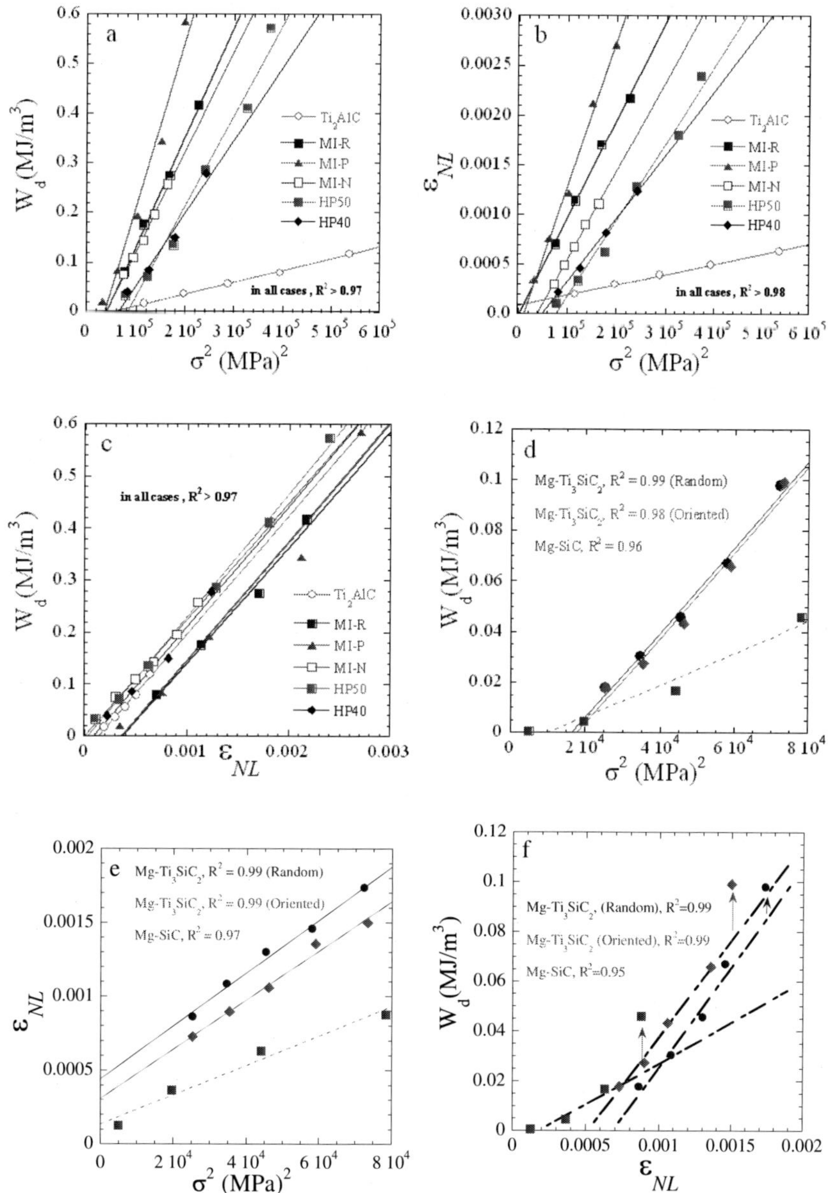

Figure 25. Plots of, a) W_d vs. σ^2, b) ε_{NL} vs. σ^2, and c) W_d vs. ε_{NL} for all Mg-Ti_2AlC composites tested in this work and those of fully dense Ti_2AlC. Plots of, d) W_d vs. σ^2, e) ε_{NL} vs. σ^2, and, f) W_d vs. ε_{NL} for Mg-Ti_3SiC_2 and Mg-SiC composites.

The main role of the nc-Mg matrix is to increase the strength of the composite, which in turn greatly enhances W_d, since the latter scales with σ^2 (Eq. 5). The nc-Mg matrix is, *however*, still soft enough to allow the Ti_2AlC grains to kink. Along the same lines, the influence of the Mg-matrix here is somewhat the opposite of the small equiaxed grains in the $Ti_2Al(C_{0.5},N_{0.5})$ solid solutions, wherein the "hard" small grains appear to constrain the majority grains from kinking [101].

The fact that at $\sigma < 420$ MPa, the HP40 porous sample dissipates more energy than the dense HP50 sample on an absolute scale (Figure 25), is in line with previous results [26]. This observation, together with the fact that the MI-P composites produce the largest loops yet, is compelling evidence that what is observed is most likely due to IKBs because, as noted previously [26], it essentially eliminates deformation mechanisms that scale with the volume of the material, and/or depend on shear alone, such as dislocation pileups. Said otherwise, had dislocation pileups been responsible for the loops, the W_d values would have probably been expected to be highest for the random, fully dense, microstructure. This conclusion cannot be overemphasized.

In contrast to Mg-Ti_2AlC samples, the W_d vs. σ^2 plots of the randomly oriented Mg-312 composite and those of the oriented Mg-312 sample (Figure 25) seem to be, within experimental scatter, identical, although the XRD results showed that the Mg-312 sample was highly oriented too. The reason for this state of affairs is not *entirely* clear at this point but can be related to the equiaxed morphology of the Ti_3SiC_2 grains that are relatively less amenable to kinking as compared to the plate-like grains of Ti_2AlC. This is best manifested by comparing the OM micrographs of Figure 14. As shown in Figure 25, for the Mg-312 and Mg-SiC composites, the linear curves of ε_{NL} vs. σ^2 plots extrapolate back to a finite ε_{NL} at zero applied loads. Apparently, these linear plots fail at low applied stresses in these composite samples, while not in the Mg-Ti_2AlC composites.

The results shown in Table 5 are important for several reasons. First, the fact that the values of Ω/b – calculated from Eqs. 5 and 6, and listed in columns 4 and 5 in Table 5, respectively – are almost identical in all cases is, as noted above, strong evidence that the micromechanism that is causing the strain nonlinearity is the same as that resulting in W_d. Hence, for example, we can exclude microcracking as a possible mechanism for W_d. Second, at 37.7±0.5 MPa, the Ω/b values obtained here for both the Mg-Ti_2AlC composites, as well as the bulk Ti_2AlC, are quite comparable. This is important because it implies that most of the energy dissipated is occurring in the Ti_2AlC phase. These values are ≈ 50 % larger than those reported previously for Ti_2AlC [26, 101]. The reason(s) for this discrepancy is most probably the differences in grain size. We have recently shown that at least for the $Ti_{n+1}AlX_n$ phases the CRSS follows a Hall-Petch type relationship [101]; this is further corroborated by Figure 26 in which Ω/b is plotted as a function of $\frac{1}{\sqrt{2\alpha}}$ for the average of Ω/b values obtained from Mg-Ti_2AlC composites and fully dense Ti_2AlC samples. Those of the fully dense and 10 vol. % porous Ti_2AlC [26] are also included. A least squares analysis of the data results in an $R^2 = 0.99$. Note that Ω/b is a function of grain size in Ti_3SiC_2 as well [18].

In contrast to the aforementioned case where the Ti_2AlC phase is responsible for most of the energy dissipated per cycle, the situation for the Mg-312 composites is substantially different. At 16 MPa, the Ω/b value appears to be an average of that of Ti_3SiC_2 (32 MPa obtained here and 30 MPa reported in [101]) and Mg (3 MPa [35, 103]). This suggest that both of the Mg and Ti_3SiC_2 contribute to Ω/b, and hence W_d.

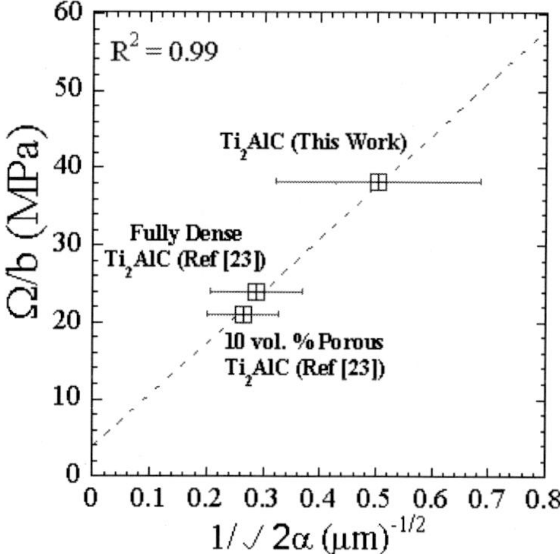

Figure 26. Plot of Ω/b as a function of $\frac{1}{\sqrt{2\alpha}}$ for the average Ω/b values obtained and those reported in [26] for fully dense and 10 vol. % porous Ti_2AlC.

Indirectly confirming this notion is the fact that Ω/b for Mg-SiC composite (5.5±0.1 MPa) is the lowest value obtained in this work and in good agreement with the 3-4 MPa reported for pure Mg [34, 35]. This implies that SiC does not contribute to the strain nonlinearities or W_d's observed. Note that to obtain the 5.5 MPa value, the last datum point in Figure 25 f, was not included for reasons discussed elsewhere, but related to the breakdown of the model in Mg at higher strains [34, 35].

To obtain the aforementioned Ω/b values, k_1 was assumed to be 2 in all cases. This result is somewhat surprising, since we expected k_1 to be texture dependent. Why that is the case is not entirely clear, but suggests that if k_1 is a function of texture, that dependence is weak.

At ≈ 1.3×10^{14} to 1.7×10^{14} m^{-2}, the values of ρ_{rev}, at comparable stress levels listed in the last column of Table 3 for all Mg-Ti_2AlC composites tested here fall in a very narrow range despite the large differences in size and shape of the original loops from which these values were extracted. Recall that ρ_{rev} is *not* the dislocation density in the sample when the load is removed, but rather the one due solely to the IKBs, i.e. ρ_{rev} given by Eq. 8. The values of ρ_{rev} fall in a narrow range despite the fact that: i) the maximum applied stresses vary in some cases by a factor of 2, ii) the N_K values vary by ~ 3 orders of magnitude, and, iii) the variations in σ_t and 2α. The values of ρ_{rev} at the maximum (and comparable) stress levels (last column in Table 5) for Mg-SiC and the Mg-312 composites also fall in the narrow range of 1.7×10^{13} to 5.7×10^{13} m^{-2}.

The same is true for Ti_3AlC_2, Ti_2AlC, $Ti_3Al(C_{0.5},N_{0.5})_2$ and $Ti_2Al(C_{0.5},N_{0.5})$ [48] where it was shown that despite large variations in the shapes and sizes of the hysteretic loops, ρ_{rev} varied by less than one order of magnitude (1×10^{13} to 9×10^{13} m^{-2}). The same is true for Mg [34] wherein N_k varied by almost 3 orders of magnitude, the *reversible* dislocation density, ρ_{rev}, varied by a factor of 3. Although not clearly understood at this point, the results of this work, and those of Refs. [34, 101] suggest that a near-constant ρ_{rev} exists to which all systems migrate, regardless of their chemistry and/or microstructure.

Lastly, the choice of the value of w = 5b, needs to be addressed. The minimum value of w is b which cannot be correct since, from Eq. 1, results in 2α values in the order of ≈ 20 μm for the Mg-Ti_2AlC composites and ≈ 35 μm in Mg-Ti_3SiC_2 composites that are much larger than the experimentally measured 2α values. Recall, 2α is the thickness of the Ti_2AlC and Ti_3SiC_2 grains along the *c*-axis, which, are estimated to be ~ 5±3μm and ~ 8±2 μm, respectively. On the other extreme, assuming w = 20b, yields 2α values ≈ 1 μm for the Mg-Ti_2AlC composites and ≈ 2 μm in Mg-Ti_3SiC_2 composites, values that are again inconsistent with the OM micrographs. Assuming w = 5b results in 2α values of the *average* Ti_2AlC and Ti_3SiC_2 grains in their composites to be ≈ 3 μm and ≈ 7 μm, respectively, results in values that are in reasonable agreement with experimentally measured 2α values of bulk Ti_2AlC and Ti_3SiC_2 grains mentioned above.

Ultimate Compressive Strength

Typically the addition of soft metallic phases to binary carbides and nitrides decreases the strength of the composites [72]. The UCS of the HP and MI composites were measured to be 800±25 and 700±10, respectively. These values are slightly lower than the 865±55 MPa, of fully dense Ti_2AlC, and are remarkably high for a 50 vol. %, essentially pure, Mg matrix composite. The reason for these extreme values is most likely attributable to the nano-grains of the Mg matrix. This is best evidenced by comparing the UCSs of the Ti_2AlC reinforced composites with those reinforced with SiC or Ti_3SiC_2, in which case the Mg-grains were *not* in the nanometer scale. At 500±25 and 460±10 MPa, respectively, their UCSs were significantly lower than the Mg-Ti_2AlC composites with comparable volume fractions. These values were also less than *half* the strengths of bulk Ti_3SiC_2. Said otherwise, when the Mg-matrix grains were *not* in the nanometer scale, the UCS values obtained were significantly lower than those obtained from their monolithic binary or ternary carbide counterparts.

The UCS values achieved here – e.g. in the HP composites – by the addition of 50 vol. % commercially pure Mg to Ti_2AlC – are , as far as we are aware, the highest ever reported for a pure Mg-matrix composite with 50 vol.% Mg. The lower UCS of the HP40 composite compared to fully dense HP50 is due to the porosity in the former. It is important to note that despite the non-linear IKB induced strains and also the permanent plastic strains (Figures 23 b and c) observed, the samples failed by shear banding at 45° to the loading axis. Given the limited number of slip systems in Mg, Ti_2AlC and Ti_3SiC_2 and the high UCS, this is not too surprising. These findings are consistent with those reported elsewhere in other Mg-matrix composites reinforced with *large* reinforcement particles [104, 105]. Interestingly, when sub-micron ceramic particles such as nano-Al_2O_3 [106-108] and nanosized Y_2O_3 particulates [109, 110] are used, the fracture behavior of the Mg matrix changed from brittle to ductile. It would be therefore worthwhile to try and make and test composites in which both the Mg-matrix *and* the reinforcing phases are both at the nanoscale.

OFFSET YIELD STRENGTH

As shown in the monotonic stress-strain plots to the point of fracture in HP50 and MI50 samples (Figures 23 b and c), the 0.2% offset yield strength (Y_S) of HP50 and MI50 composites were measured to be ~ 690 and 490 MPa, respectively. The irreversible permanent plastic strains were also measured to be ~ 2.2 and 4.6%, respectively. The reason for these differences between the two microstructures can, most likely, be attributed to the finer grain size of the nanocrystalline Mg matrix in the HP50 sample. These results are directly corroborated by the microhardness results shown below (see section 4.4.5).

EFFECTIVE YOUNG'S MODULI

As shown in Table 3, $\overline{E}$ is a function of kinking and depends on the size and extent of the hysteresis stress-strain curves. The MI-N composites exhibited slightly higher $\overline{E}$ values compared to MI-P. The $\overline{E}$'s of the Mg-Ti_3SiC_2 composites seem to be close to the Mg-Ti_2AlC ones, despite the fact that Ti_3SiC_2 is slightly stiffer than Ti_2AlC (~ 343 vs. 277 GPa reported in Ref. [111], respectively). The fact that $\overline{E}$ in their corresponding composites is most likely due to the lack of kinking of the Mg nanograins in the Mg-Ti_2AlC system and their kinking in the Mg-312 composites. The Mg-SiC composite, on the other hand, exhibited the largest $\overline{E}$ values believed to be due to the higher modulus of elasticity of SiC (~ 475 GPa [102]) and the small size of the stress-strain loops associated with Mg-SiC composite.

With the exception of MI-P sample, the average $\overline{E}$ values for Mg-Ti_2AlC, Mg-Ti_3SiC_2 and Mg-SiC composites (listed in Table 3), within experimental scatter, fall in between the rule of mixtures' upper and lower bounds given by $E_c(u) = E_m V_m + E_p V_p$ and $E_c(l) = E_m E_p / (V_m E_p + V_p E_m)$, respectively [112], wherein $E_c(u)$ and $E_c(l)$ are the composites' moduli obtained from upper and lower bounds; E_m, E_P, V_m and V_P are the moduli and volume fractions of the Mg matrix and reinforcement particles, respectively. It follows that, G and ν of the Mg-Ti_2AlC, Mg-Ti_3SiC_2 and Mg-SiC composites are assumed to be ~ the averages of rule-of-mixtures' lower and upper bounds [112] of the corresponding composites. All values are listed in Table 3. The G and ν values of Ti_2AlC [26], Ti_3SiC_2 [101], SiC [102], and Mg [35] are *118* GPa and *0.2*, *144* GPa and *0.2*, *192* and *0.142*, and *19* GPa and *0.35*, respectively.

VICKERS MICROHARDNESS, V_H

At 2.0±0.1 and 1.5±0.1 GPa, the V_H obtained for HP and MI composites are again remarkably high for a 50 vol% Mg composite. The hardness enhancement in the HP sample compared to its MI counterpart can be attributed to the smaller nc-Mg matrix in the former. This is also directly corroborated by the higher Y_S and lower permanent plastic strains measured for the HP sample.

The MI-N orientation is also ~ 25 % harder than its counterpart, MI-P. Since both samples were obtained from the same billet, it is fair to conclude that the orientation of the basal planes is responsible for the difference. Image analysis of the MI-P and MI-N indented surfaces, showed that the area covered by the "harder" Ti_2AlC grains was ≈ 7±1% larger in the former than the latter. This probably, partially, explains the differences in hardness.

This is true despite the results of Kooi *et al.* [113] who showed, using a nanoindenter and orientation image microscopy, that the Berkovich hardness was lower in Ti_3SiC_2 grains that were indented perpendicular to the basal planes, i.e. when loaded along the c-direction. Kooi *et al.* performed their nano-indentations on individual Ti_3SiC_2 grains with their basal planes either oriented parallel or perpendicular to the surface. The results reported here are for Mg-Ti_2AlC composites, where the Mg-grains are responsible for most of the deformation strain, which is presumably why the volume fraction of Mg is more important than the orientation of the Ti_2AlC grains.

The V_H values of the nc Mg-Ti_2AlC composites tested are comparable to those of Mg-matrix composites in which the reinforcing phase is significantly harder. For example, at 2.0±0.1 GPa the HP50 results are comparable to those of the Mg-SiC samples, or Mg-TiC composites, with 56 vol. % TiC, reported by Contreras *et al.* [114]. This value is also higher than, i) the Mg-312 samples, ii) > ~ 1.1 GPa values reported in Mg-TiC composites fabricated by powder metallurgy [115], iii) the ~ 1 GPa values reported for Mg-alloy matrix composites reinforced with TiC [116]. In all cases, the large hardness differences between SiC (V_H ~ 28 GPa [102]), TiC (V_H ~ 35 GPa [102, 117]) and Ti_3SiC_2 (steady state V_H = 4.0±1.0 GPa) [11], on the one hand, and Ti_2AlC (steady state V_H = 3.0±0.1 GPa), on the other, is compensated by the nano-crystalline nature of the Mg-matrix in the Mg-Ti_2AlC composites.

Like most MAX phases [25, 100, 118], the hardness values of monolithic Ti_2AlC are initially high, decrease with increasing load, and then asymptote at higher loads. Interestingly, the composites' hardness values are not a function of load and fall in between those of pure Mg and monolithic Ti_2AlC. Like the vast majority of MAX phases [7, 25, 118], no cracks are observed to emanate from the corners of the Vickers indentations of the composite samples. This damage tolerance is a hallmark of the MAX phases and results from the activation of basal slip which allows the material to absorb energy locally by various energy absorbing mechanisms such as microcracking, delamination, grain buckling and grain pull-out [7, 25]. The plastic deformation of the Mg matrix must also play an important role. The desirability of such high damage tolerance in potential applications cannot be overemphasized.

CONCLUSION

MAXMETs are all KNE solids characterized by the formation of fully reversible hysteretic stress-strain loops under uniaxial cyclic compression. The microscale model developed to analyze and explain kinking nonlinear elasticity in KNE solids is in excellent agreement with the experimental results obtained in the MAXMET composites and their monolithic counterparts. These findings are important when designing solids with ultrahigh damping capabilities at high stresses and decent elastic moduli. Depending on the application, and the stress levels required during service, different MAX-Mg composites can be used. For relatively high stress applications, the basal planes of the MAX phases should be loaded edge-on to yield the highest W_d values.

The Ω/b values obtained from the model are a function of their constituents and whether or not they kink. Because for the Mg-Ti_2AlC composites, the Ω/b values are almost identical to those of bulk Ti_2AlC, it is reasonable to assume that the latter is doing most of the kinking. In contradistinction, because the Mg matrix of the Mg-312 composites are *not* at the nanoscale and are thus more prone to kinking, the Ω/b values obtained are considerably less than those of the Mg-Ti_2AlC system, and seem to be the average of the Ω/b values of Mg and Ti_3SiC_2. The Mg-SiC, in which only the Mg matrix is presumably *kinking*, has the lowest Ω/b values that are in line with those of pure Mg.

The ρ_{rev} values fall in a narrow range, suggesting that an equilibrium state, to which all the systems migrate, exists.

Despite a 50 vol.% loading of essentially pure Mg, the highest UCS and V_H values of Mg-Ti_2AlC composites were 800±25 MPa and 2.0±0.1 GPa. These extreme values are related to the nano-size of the Mg grains. The fact that these composites are also readily fabricated by MI, machinable, light-weight and stiff, as well as highly damping, should render them useful materials for a host of applications.

REFERENCES

[1] Sundberg, M., et al., *Alumina Forming High Temperature Silicides and Carbides.* Ceramics International, 2004. 30: p. 1899-1904.

[2] Barsoum, M.W., *Physical Properties of the MAX Phases*, in *Encyclopedia of Materials Science and Technology*, R.W.C. K. H. J. Buschow, M. C. Flemings, E. J. Kramer, S. Mahajan and P. Veyssiere, Editor. 2006, Elsevier: Amsterdam.

[3] Barsoum, M.W., D. Brodkin, and T. El-Raghy, *Layered Machinable Ceramics For High Temperature Applications.* Scrip. Met. et. Mater., 1997. 36: p. 535-541.

[4] Barsoum, M.W. and T. El-Raghy, *Synthesis and Characterization of a Remarkable Ceramic: Ti_3SiC_2.* J. Amer. Cer. Soc., 1996. 79(7): p. 1953-1956.

[5] Barsoum, M.W. and T. El-Raghy, *Room Temperature Ductile Carbides.* Metallurgical and Materials Trans., 1999. 30A: p. 363-369.

[6] Barsoum, M.W., et al., *Thermal Properties of Ti_3SiC_2.* J. Phys. Chem. Solids, 1999. 60: p. 429.

[7] El-Raghy, T., et al., *Damage Mechanisms Around Hardness Indentations in Ti_3SiC_2.* J. Amer. Cer. Soc., 1997. 80: p. 513-516.

[8] Barsoum, M.W., T. El-Raghy, and L. Ogbuji, *Oxidation Behavior of Ti_3SiC_2 in the Temperature Range of 900–1400°C.* J. Electrochem. Soc., 1997. 144: p. 2508–16.

[9] Barsoum, M.W. and T. El-Raghy, *A Progress Report on Ti_3SiC_2, Ti_3GeC_2 and the H-Phases, M_2BX.* J. Mater. Synth. Process., 1997. 5: p. 197–216.

[10] Barsoum, M.W., G. Yaroschuck, and S. Tyagi, *Fabrication and Characterization of M_2SnC (M = Ti, Zr, Hf and Nb).* Scr. Mater, 1997. 10: p. 1583–91.

[11] Low, I.M., et al., *Contact Damage Accumulation in Ti_3SiC_2.* ,J. Amer. Cer. Soc. , 1998. 81: p. 225-28.

[12] Farber, L., et al., *Dislocations and Stacking Faults in Ti_3SiC_2*. J. Am. Ceram. Soc., 1998. 81(6): p. 1677–81.

[13] El-Raghy, T. and M.W. Barsoum, *Processing and mechanical properties of Ti_3SiC_2: part I: reaction path and microstructure evolution.* . J. Amer. Cer. Soc., 1999. 82: p. 2849-54

[14] El-Raghy, T., et al., *Processing and mechanical properties of Ti_3SiC_2: II, effect of grain size and deformation temperature.* J. Amer. Cer. Soc., 1999. 82 p. 2855-2860.

[15] Barsoum, M.W., et al., *Dislocations, Kink Bands and Room Temperature Plasticity of Ti_3SiC_2*. Met. Mater. Trans., 1999. 30A: p. 1727-1738

[16] Barsoum, M.W., *The $M_{N+1}AX_N$ Phases: a New Class of Solids; Thermodynamically Stable Nanolaminates.* Prog. Solid State Chem, 2000. 28: p. 201-281.

[17] Wang, X.H. and Y.C. Zhou, *Oxidation behavior of Ti_3AlC_2 at 1000-1400 °C in air.* Corrosion Science, 2003. 45: p. 891-907.

[18] Barsoum, M.W., et al., *Microscale Modeling of Kinking Nonlinear Elastic Solids.* Phys. Rev. B., 2005. 71: p. 134101.

[19] Murugaiah, A., et al., *Spherical Nanoindentations in Ti_3SiC_2*. J. Mater. Res., 2004. 19: p. 1139-1148

[20] Barsoum, M.W., et al., *Dynamic Elastic Hysteretic Solids and Dislocations.* Phys. Rev. Lett., 2005. 94: p. 085501.

[21] Barsoum, M.W., et al., *Kink Bands, Nonlinear Elasticity and Nanoindentations in Graphite.* Carbon, 2004. 42: p. 1435-1445.

[22] Barsoum, M.W., et al., *Kinking Nonlinear Elastic Solids, Nanoindentations and Geology.* Phys. Rev. Lett., 2004. 92: p. 255508-1.

[23] Barsoum, M.W., et al., *Fully Reversible, Dislocation-Based Compressive Deformation of Ti_3SiC_2 to 1 GPa.* Nature Materials, 2003. 2: p. 107-111.

[24] Orowan, E., *A Type of Plastic Deformation New In Metals.* Nature, 1942. 149: p. 463-464.

[25] Amini, S., et al., *Synthesis and elastic and mechanical properties of Cr2GeC.* Journal of Materials Research, 2008. 23(8): p. 2157-2165.

[26] Zhou, A.G., et al., *Incipient and Regular Kink Bands in Dense and Porous Ti_2AlC.* Acta Mater., 2006. 54: p. 1631-1639.

[27] Barsoum, S.B.A.Z.M.W., *On Nanoindentations, Kinking Nonlinear Elasticity of Mica Single Crystals and Their Geological Implications.* In Print.

[28] Basu, S., M.W. Barsoum, and S.R. Kalidindi, *Sapphire: A kinking nonlinear elastic solid.* J. App. Phys., 2006. 99: p. 063501.

[29] Basu, S. and M.W. Barsoum, *Deformation micromechanisms of ZnO single crystals as determined from spherical nanoindentation stress-strain curves.* Journal of Materials Research, 2007. 22(9): p. 2470-7.

[30] Basu, S., et al., *Spherical nanoindentation and deformation mechanisms in freestanding GaN films.* Journal of Applied Physics, 2007. 101(8).

[31] Basu, S., A. Zhou, and M.W. Barsoum, *Reversible Dislocation Motion under Contact Loading in LiNbO3 Single Crystal.* Journal of Materials Research, Accepted.

[32] Zhou, A.G., S. Basu, and M.W. Barsoum, *Kinking nonlinear elasticity, damping and microyielding of hexagonal close-packed metals.* Acta Materialia, 2008. 56(1): p. 60-67.

[33] Zhou, A., *Kinking Nonlinear Elasticity: Theory and Experiments.* Ph.D. Thesis, 2008. Drexel University

[34] Zhou, A. and M. Barsoum, *Kinking Nonlinear Elasticity and the Deformation of Magnesium.* Metallurgical and Materials Transactions A, 2009.

[35] Zhou, A.G., S. Basu, and M.W. Barsoum, *Kinking nonlinear elasticity, damping and microyielding of hexagonal close-packed metals.* Acta Materialia, 2008. 56(1): p. 60-7.

[36] Zhou, A.G., et al., *On The Kinking Nonlinear Elastic Deformation of Cobalt.* Submitted for Publication.

[37] Amini, S., C. Ni, and M.W. Barsoum, *Processing, microstructural characterization and mechanical properties of a Ti_2AlC/nanocrystalline Mg-matrix composite.* Composites Science and Technology, 2009. 69: p. 414-420.

[38] Amini, S. and M.W. Barsoum, *On the Effect of Texture on the Mechanical Properties of Nano-crystalline Mg-Matrix Composites Reinforced with MAX Phases.* Submitted for Publication, 2009.

[39] Basu, S., M.W. Barsoum, and S.R. Kalidindi, *Sapphire: A kinking nonlinear elastic solid.* Journal Of Applied Physics, 2006. 99(6).

[40] Frank, F.C. and A.N. Stroh, *On the Theory of Kinking.* Proc. Phys. Soc., 1952. 65: p. 811-821

[41] Basu, S., A. Zhou, and M.W. Barsoum, *Reversible dislocation motion under contact loading in LiNbO3 single crystal.* Journal of Materials Research, 2008. 23(5): p. 1334-1338.

[42] Barsoum, M.W. and M. Radovic, *Mechanical Properties of the MAX Phases*, in *Encyclopedia of Materials Science and Technology*, R.W.C. K. H. J. Buschow, M. C. Flemings, E. J. Kramer, S. Mahajan and P. Veyssiere, Editor. 2004, Elsevier: Amsterdam.

[43] Amini, S. and M.W. Barsoum, *On the effect of texture on the mechanical and damping properties of nanocrystalline Mg-matrix composites reinforced with MAX phases.* Materials Science and Engineering a-Structural Materials Properties Microstructure and Processing, 2010. 527(16-17): p. 3707-3718.

[44] Amini, S., C. Ni, and M.W. Barsoum, *Processing, microstructural characterization and mechanical properties of a Ti2AlC/nanocrystalline Mg-matrix composite.* Composites Science and Technology, 2009. 69(3-4): p. 414-420.

[45] Amini, S., et al., *Synthesis and elastic and mechanical properties of Cr2GeC.* Journal of Materials Research, 2008. 23(8): p. 2157-2165.

[46] A. Zhou, D.B., S. Vogel, O. Yeheskel and M. W. Barsoum, *On the Kinking Nonlinear Elastic Deformation of Polycrystalline Cobalt,.* Materials Science and Engineering: A. In Press, Accepted Manuscript.

[47] Basu, S., A. Zhou, and M.W. Barsoum, *On spherical nanoindentations, kinking nonlinear elasticity of mica single crystals and their geological implications.* Journal of Structural Geology, 2009. 31(8): p. 791-801.

[48] Zhou, A.G. and M.W. Barsoum, *Kinking nonlinear elastic deformation of Ti3AlC2, Ti2AlC, Ti3Al(C-0.5,N-0.5)(2) and Ti2Al(C-0.5,N-0.5).* Journal of Alloys and Compounds, 2010. 498(1): p. 62-70.

[49] Amini, S., et al., *Synthesis and Elastic and Mechanical Properties of Cr2GeC.* Journal of Materials Science, 2008. In Print.

[50] Westengen, H. and P. Bakke, *Magnesium die casting alloys for use in applications exposed to elevated temperatures: can they compete with aluminium?* Materials Science Forum, 2003. 419-422: p. 35-40.

[51] Ye, H.Z. and L. Xing Yang, *Review of recent studies in magnesium matrix composites.* Journal of Materials Science, 2004. 39(20): p. 6153-71.

[52] Luo, A., et al., *Magnesium castings for automotive applications.* JOM, 1995. 47(7): p. 28-31.

[53] Clow, B.B., *Magnesium industry review.* Advanced Materials & Processes, 1996. 150(4): p. 2.

[54] Lazan, B.J., *Damping of materials and members in structural mechanics.* Pergamon Press, Oxford, New York, 1968.

[55] Capel, H., et al. *Correlation between manufacturing conditions and properties of carbon fibre reinforced Mg.* 2000. London, UK: Inst. Mater.

[56] Chua, B.W., L. Lu, and M.O. Lai, *Influence of SiC particles on mechanical properties of Mg based composite.* Composite Structures, 1999. 47(1): p. 595-601.

[57] Li, L., et al., *Improvement of microstructure and mechanical properties of AZ91/SiC composite by mechanical alloying.* Journal of Materials Science, 2000. 35(22): p. 5553-5561.

[58] Badini, C., et al., *Precipitation phenomena in B4C-reinforced magnesium-based composite.* Materials Science & Engineering A (Structural Materials: Properties, Microstructure and Processing), 1992. A157(1): p. 53-61.

[59] Mikucki, B.A., W.E. Mercer, II, and W.G. Green, *Extruded magnesium alloys reinforced with ceramic particles.* Light Metal Age, 1990. 48(5-6): p. 12.

[60] Luo, A., *Processing, microstructure, and mechanical behavior of cast magnesium metal matrix composites.* Metallurgical and Materials Transactions A: Physical Metallurgy and Materials Science, 1995. 26A(9): p. 2445-2455.

[61] Saravanan, R.A. and M.K. Surappa, *Fabrication and characterisation of pure magnesium-30 vol.% SiCP particle composite.* Materials Science & Engineering A (Structural Materials: Properties, Microstructure and Processing), 2000. A276(1-2): p. 108-16.

[62] Lim, S.-w. and T. Choh, *Effect of alloying elements on SiC particulate dispersion behavior in molten magnesium.* Nippon Kinzoku Gakkaishi/Journal of the Japan Institute of Metals, 1992. 56(2): p. 210-217.

[63] Purazrang, K., K.U. Kainer, and B.L. Mordike, *Fracture toughness behaviour of a magnesium alloy metal-matrix composite produced by the infiltration technique.* Composites, 1991. 22(6): p. 456-462.

[64] Wu, K., et al., *Crystallographic orientation relationship between SiCw and Mg in squeeze-cast SiCw/Mg composites.* Journal of Materials Science Letters, 1999. 18(16): p. 1301-1303.

[65] Zheng, M., et al., *Interfacial bond between SiCw and Mg in squeeze cast SiCw/Mg composites.* Materials Letters, 1999. 41(2): p. 57-62.

[66] Wu, K., et al., *Interfacial reaction in squeeze cast SiCw/AZ91 magnesium alloy composite.* Scripta Materialia, 1996. 35(4): p. 529-534.

[67] Zheng, M., et al., *Precipitates in aged SiCw/AZ91 magnesium matrix composite.* Journal of Materials Science Letters, 1997. 16(13): p. 1106-1108.

[68] Kevorkijan, V., *Mg AZ80/SiC composite bars fabricated by infiltration of porous ceramic preforms.* Metallurgical and Materials Transactions a-Physical Metallurgy and Materials Science, 2004. 35A(2): p. 707-715.

[69] Dong, Q., et al., *Synthesis of TiCp reinforced magnesium matrix composites by in situ reactive infiltration process.* Materials Letters, 2004. 58(6): p. 920-926.

[70] Kaneda, H. and T. Choh, *Fabrication of particulate reinforced magnesium composites by applying a spontaneous infiltration phenomenon.* Journal of Materials Science, 1997. 32(1): p. 47-56.

[71] Kevorkijan, V., T. Kosmac, and K. Kristoffer, *Spontaneous reactive infiltration of porous ceramic preforms with Al-Mg and Mg in the presence of both magnesium and nitrogen - New experimental evidence.* Materials and Manufacturing Processes, 2002. 17(3): p. 307-322.

[72] *ASM Handbooks Online*. 2008, ASM Intrnational.

[73] Barsoum, M.W., M. Ali, and T. El-Raghy, *Processing and characterization of Ti_2AlC, Ti_2AlCN and $Ti_2AlC_{0.5}N_{0.5}$.* Met Mater Trans, 2000. 31A: p. 1857-65.

[74] Gupta, S., et al., *Ta₂AlC and Cr₂AlC Ag-based composites-New solid lubricant materials for use over a wide temperature range against Ni-based superalloys and alumina.* Wear, 2007. 262(11-12): p. 1479-1489.

[75] Zhang, Z. and S. Xu, *Copper-Ti_3SiC_2 composite powder prepared by electroless plating under ultrasonic environment.* Rare Metals, 2007. 26(4): p. 359-364.

[76] Ngai, T.L., et al., *Studies on preparation of Ti_3SiC_2 particulate reinforced Cu matrix composite by warm compaction and its tribological behavior.* Materials Science Forum, 2007. 534-536: p. 929-32.

[77] Ngai, T.L., L. Yuanyuan, and Z. Zhaoyao, *A study on Ti_3SiC_2 reinforced copper matrix composite by warm compaction powder metallurgy.* Materials Science Forum, 2007. 532-533: p. 596-9.

[78] Barsoum, M.W., et al., *Long Time Oxidation Study of Ti_3SiC_2, Ti_3SiC_2/SiC and Ti_3SiC_2/TiC Composites in Air.* J. Electrochem. Soc., 2003. 150(4): p. B166-B175.

[79] Ho-Duc, L.H., T. El-Raghy, and M.W. Barsoum, *Synthesis and characterization of 0.3VfTiC-TiSiC2 and 0.3VfSiC Ti3SiC2 composites.* J. of Alloys and Compounds, 2003. 350: p. 303-312.

[80] Wu, J., Y. Zhou, and C. Yan, *Mechanical and electrical properties of Ti2SnC dispersion-strengthened copper.* Zeitschrift fuer Metallkunde/Materials Research and Advanced Techniques, 2005. 96(8): p. 847-852.

[81] Kainer, K.U., *Magnesium alloys and technology*. 2003: Wiley-VCH.

[82] Decker., R.F., *The renaissance of magnesium.* Advanced Materials & Processes, 1998. 154(3): p. 31-33

[83] Stalmann, A., et al., *Properties and processing of magnesium wrought products for automotive applications.* Advanced Engineering Materials, 2001. 3(12): p. 969-974.

[84] Mortensen, A., K. Anthony, and Z. Carl, *Melt Infiltration of Metal Matrix Composites*, in *Comprehensive Composite Materials*. 2000, Pergamon: Oxford. p. 521-554.

[85] Fraczkiewicz, M., A.G. Zhou, and M.W. Barsoum, *Mechanical Damping in Porous Ti_3SiC_2.* Acta Mater., 2006. In press.

[86] Hull, D., *Introduction to Dislocation*. 1965: Oxford: Pergamon Press.

[87] Barsoum, M.W., et al., *Kinking Nonlinear Elastic Solids*, in *Encyclopedia of Materials: Science and Technology*. 2010, Elsevier Science: Amsterdam.

[88] Reed-Hill, R.E., E.P. Dahlberg, and W.A. Slippy, *Some Anelastic Effects in Zirconium at Room Temperature Resulting from Prestrain at 77K.* Trans. Metallurgical Soc. of AIME., 1965. 233: p. 1766-1771.

[89] Zhou, A.G. and M.W. Barsoum, *Kinking Nonlinear Elastic Deformation of Ti_3AlC_2, Ti_2AlC, $Ti_3Al(C_{0.5},N_{0.5})_2$ and $Ti_2Al(C_{0.5},N_{0.5})$.* Submitted for Publication, 2009.

[90] Murugaiah, A., et al., *Tape Casting, Pressureless Sintering and Grain Growth in Ti_3SiC_2 Compacts.* J. Amer. Cer. Soc., 2004. 87: p. 550-556

[91] Jeitschko, W., H. Nowotny, and F. Benesovsky, *Kohlenstoffhaltige ternare Verbindungen (H-Phase).* Monatsh. Chem., 1963. 94: p. 672.

[92] Kisi, E.H., et al., *Structure and Crystal Chemistry of Ti_3SiC_2.* J. Phys. and Chem. of Solids, 1998. 59(9): p. 1437-1442.

[93] Cullity, B.D., *Elements of x-ray diffraction.* 1978.

[94] E. Brandes, G.B.B., c. Smithells, *Smithells Metals Reference Book.* 1998: Oxford; Boston: Butterworth-Heinemann.

[95] Amini, S., et al., *On the Stability of Mg Nanograins to Coarsening after Repeated Melting.* Nano Letters, 2009. 9(8): p. 3082-3086.

[96] Myhra, S., A. Crossley, and M.W. Barsoum, *Crystal chemistry from XPS analysis of carbide-derived Mn+1AXn (n=1) nano-laminate compounds.* J. Phys. and Chem. of Solids, 2002: p. 811-817.

[97] Barsoum, M.W., *The Mn+1AXn Phases and Their Properties.* Ceramics Science and Technology, 2009. 2.

[98] Barsoum, M.W., et al., *The Topotaxial Transformation of Ti_3SiC_2 To Form a Partially Ordered Cubic $TiC_{0.67}$ Phase by the Diffusion of Si into Molten Cryolite.* J. Electrochem. Soc. , 1999. 146(3919-3923).

[99] El-Raghy, T., M.W. Barsoum, and M. Sika, *Reaction of Al with Ti_3SiC_2 in the 800-1000 °C Temperature Range.* Mater. Sci. Eng. A, 2001. 298: p. 174.

[100] Amini, S., M.W. Barsoum, and T. El-Raghy, *Synthesis and mechanical properties of fully dense Ti_2SC.* Journal of the American Ceramic Society, 2007. 90(12): p. 3953-3958.

[101] Barsoum, A.G.Z.M.W., *Kinking nonlinear elastic deformation of Ti3AlC2, Ti2AlC, Ti3Al(C0.5,N0.5)2 and Ti2Al(C0.5,N0.5).* Submitted for Publication, 2009.

[102] Pierson, H.O., *Handbook of Refractory Carbides and Nitrides.* 1996, Westwood, NJ: Noyes Publications.

[103] Zhou, A.G. and M.W. Barsoum, *Kinking nonlinear elasticity and the deformation of Mg.* METALLURGICAL AND MATERIALS TRANSACTIONS A, In print.

[104] Lim, S., M. Gupta, and L. Lu, *Processing, microstructure, and properties of Mg-SiC composites synthesised using fluxless casting process.* Materials Science and Technology, 2001. 17(7): p. 823-32.

[105] Hassan, S.F. and M. Gupta, *Development of ductile magnesium composite materials using titanium as reinforcement.* Journal of Alloys and Compounds, 2002. 345(1-2): p. 246-251.

[106] Hassan, S.F. and M. Gupta, *Development of high performance magnesium nano-composites using nano-Al_2O_3 as reinforcement.* Materials Science and Engineering A, 2005. 392(1-2): p. 163-168.

[107] Hassan, S.F., M.J. Tan, and M. Gupta, *High-temperature tensile properties of Mg/Al_2O_3 nanocomposite.* Materials Science and Engineering A, 2008. 486(1-2): p. 56-62.

[108] Hassan, S.F. and M. Gupta, *Effect of submicron size* Al_2O_3 *particulates on microstructural and tensile properties of elemental Mg.* Journal of Alloys and Compounds, 2008. 457(1-2): p. 244-250.

[109] Hassan, S.F. and M. Gupta, *Development and characterization of ductile Mg/Y2O3 nanocomposites.* Transactions of the ASME. Journal of Engineering Materials and Technology, 2007. 129(3): p. 462-7.

[110] Hassan, S.F. and M. Gupta, *Effect of different types of nano-size oxide particulates on microstructural and mechanical properties of elemental Mg.* Journal of Materials Science, 2006. 41(8): p. 2229-36.

[111] Radovic, M., et al., *On the elastic properties and mechanical damping of* Ti_3SiC_2, Ti_3GeC_2, $Ti_3Si_{0.5}Al_{0.5}C_2$ *and* Ti_2AlC *in the 300-1573 K temperature range.* Acta Materialia, 2006. 54(10): p. 2757-2767.

[112] Matthew, F.L., *Composite Materials: Engineering and Science*. 1994.

[113] Kooi, B.J., et al., *Ti3SiC2: A damage tolerant ceramic studied with nanoindentations and transmission electron microscopy.* Acta Materialia, 2003. 51: p. 2859-2872.

[114] Contreras, A., V.H. López, and E. Bedolla, *Mg/TiC composites manufactured by pressureless melt infiltration.* Scripta Materialia, 2004. 51(3): p. 249-253.

[115] Xiu, K., et al., *Fabrication of TiCp/Mg composites by powder metallurgy.* Journal of Materials Science, 2006. 41(5): p. 1663-6.

[116] Jiang, Q.C., et al., *Fabrication of TiCp/Mg composites by the thermal explosion synthesis reaction in molten magnesium.* Materials Letters, 2003. 57(16-17): p. 2580-2583.

[117] Kumashiro, Y., et al., *The micro-Vickers hardness of TiC single crystals up to 1500C.* Journal of Materials Science, 1977. 12(3): p. 595-601.

[118] Ganguly, A., T. Zhen, and M.W. Barsoum, *Synthesis and Mechanical Properties of* Ti_3GeC_2 *and* $Ti_3(Si_xGe_{1-x})C_2$ $(x = 0.5, 0.75)$ *Solid Solutions.* J. Alloys and Compounds, 2004. 376: p. 287-295.

In: Metal Matrix Composites
Editor: J. Paulo Davim

ISBN: 978-1-61209-771-8

Chapter 3

THERMAL EXPANSION OF AL- AND MG-MATRIX COMPOSITES WITH DIFFERENT REINFORCEMENT ARCHITECTURES

H. P. Degischer*, F. Lasagni, M. Schöbel, T. Huber[1] and M. A. Aly[2]

Vienna University of Technology,
[1]Institute of Materials Science and Technology, Karlsplatz, Vienna, Austria, Currently Siemens VAI/Austria
[2]Currently University/Egypt

ABSTRACT

The thermal expansion of different types metal matrix composites (MMC) is investigated by dilatometry. The instantaneous linear coefficients of thermal expansion (CTE) are determined for particle, interpenetrating and short fiber reinforced (SFRM) Al-Si alloy matrices between room temperature and 500°C. Particle or short fiber preforms, infiltrated by Al-Si-casting alloys, produce 3D reinforcement architectures connected by Si-bridges. The CTE(T) for unidirectional continuous fiber reinforced (CFRM) Al- and Mg-matrices are determined in different directions between -50 and 200°C. The anisotropy in thermal expansion of SFRM and CFRM is correlated to their volumetric expansion. Predictions of thermo-elastic models are only valid as long as the matrix deforms elastically, the range of which is estimated by the volume misfit between matrix and reinforcement. Deviations are explained by plastification of the matrix and temperature dependent changes in the micro-porosity of MMC with interconnected architectures of reinforcement or unidirectional CFRM. The conservation of the volume of the constituents is compared with the experimental volume variations with temperature.

*Corresponding author, E-mail: hpdegi@pop.tuwien.ac.at

1. Introduction

Metal matrix composites (MMC) have been developed to improve the mass related mechanical properties with respect to the matrix metal by embedding ceramic reinforcements [1]. High thermal conductivity and low thermal expansion are the main requirements for materials used in thermal management components for electronics [2-4] and combustion engines [5]. Ceramic reinforcements considered here exhibit significantly higher Young's moduli and smaller coefficients of thermal expansion (*CTE*) reducing the thermal expansion of the MMC, but introduce internal stresses due to the mismatch in elastic properties and thermal expansion. The *CTE* of MMC [6] can be estimated by thermo-elastic models [7,8] on the basis of the elastic constants and the *CTE* of the constituents. Those predictions fail as soon as the matrix plastifies when thermally induced stresses exceed its yield strength or induce creep. 2D simulations for elasto-plastic matrices of different contiguities with the reinforcement [8-10] predict that the thermal expansion of MMC with interconnected reinforcement is less than that for discontinuous reinforcement. A quantitative study [11] of the effect of the reinforcement's contiguity on the thermal expansion concludes that particle reinforced metals behave like interpenetrating composites when a certain portion of connectivity between the particles exist. Small volume fractions of voids originated from processing reduce the *CTE* further as determined experimentally [12] and by simulation [8,10]. Both, isotropic and anisotropic architectures of the rigid reinforcement, require a three-dimensional consideration. Deviations from the thermo-elastic behavior of particle (PRM), short fiber (SFRM) and continuous fiber (CFRM) reinforced Al and Mg matrices [13-16], were detected by dilatometry and will be considered here.

2. Methodology

2.1. Investigated Composites and Properties of the Ingredients

Table 1 lists the constituents of the investigated Al- and Mg-matrix composites and their relevant properties, of which the mismatch in *CTE* is of particular interest. Table 2 lists the tested MMC with the corresponding expansion data. The elastic and physical properties of both alumina fibers are isotropic, but the Young's moduli are different for the short fibers of the brand Saffil® and the continuous Nextel N610® fibers. The C-fibers themselves are anisotropic as presented by the longitudinal and the radial properties. The high modulus C-fiber (HM) Toray M40 exhibits a negative *CTE* in fiber direction. The relatively poor transverse properties of C-fibers are approximated [17]. The Young's moduli of Al-Si matrix alloys are slightly higher and the *CTE* are lower than the corresponding properties quoted for pure Al [15]. Al-Si matrices are tested with SiC reinforcements and with Saffil® short fibers, where the eutectic Si can be considered as an additional reinforcing phase [14,15]. The small additions of Mg to the Al-matrix and those of Al to the Mg-matrix have been employed to increase the yield strength of the matrices of CFRM [16]. The *CTE(T)* of the matrices increases considerably with temperature, whereas their yield strength decreases as indicated in Table 1. The misfit strains between matrix and reinforcement increase with temperature, whereas the elastic accommodation of the misfit decreases producing increasing matrix plastification with temperature.

In Table 1, the linear expansion misfit $\Delta\varepsilon$ for 100K temperature change around 100°C is compared for various composite systems. Mg/C-HM exhibits the highest expansion misfit in fiber direction among the selected MMC, Al/C-HM the smallest $\Delta\varepsilon$ transverse to the fibers. The yield strengths σ_y of pure Al and pure Mg are indicated, which limit their elastic straining. The maximum achievable elastic strains of the matrices are given as well. ΔT_{elast} represents the maximum temperature change to produce an expansion misfit of the freely expanding constituents, which corresponds to the elastic range of the matrix $\Delta\varepsilon_m^{elast}$: for example, increasing the temperature from RT by 25K or 60K is sufficient to reach such a linear misfit in Al/C-HM in longitudinal or transverse direction, respectively. Relatively small temperature changes cause internal stresses higher than the yield strength of the matrix.

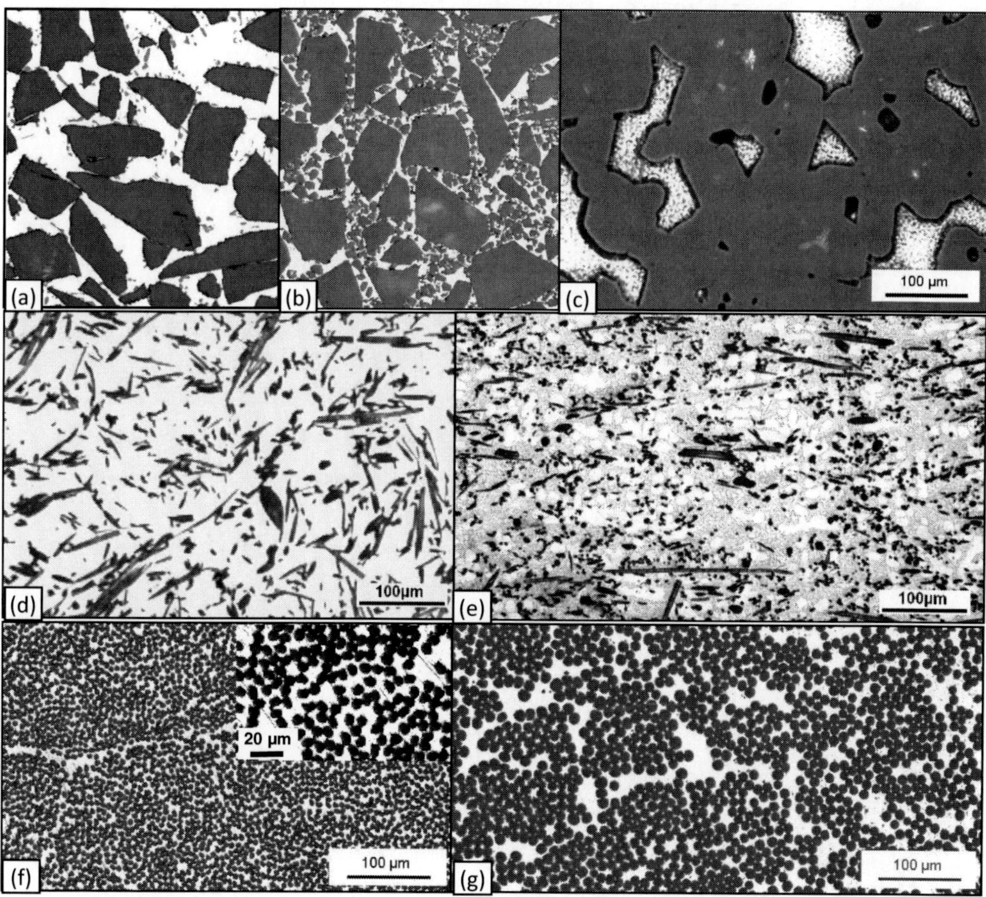

Figure 1. Light micrographs of investigated MMC: PRM: a) AlSi7Mg/SiC/55p[C], b) AlSi7Mg/SiC/70p; IMC: c) AlSi7/Mg/R-SiC/85i; SFRM: d) AlSi1/Al2O3/20s in fiber plane, e) AlSi12/Al2O3/20s transverse (fiber plane horizontal, eutectic regions in the matrix grey); CFRM in transverse sections: f) MgAlo.6/C-M40/70f-UD, enlarged inset shows touching fibers, g) AlMg1/Al2O3/65f-UD.

Table 1. Young's moduli E, linear physical thermal expansion coefficients CTE of constituents and mismatch ΔCTE (RT - 500°C); linear expansion misfit $\Delta\varepsilon$/100K and temperature interval ΔT to reach the elastic strain limit of Al- or Mg-matrices [18]

Material	E [GPa] (RT-500°C)	CTE [ppm/K] (RT-500°C)	ΔCTE [ppm/K] RT-500°C	$\Delta\varepsilon$ (50-150°C) [%]	max. ΔT_{elast} [K] for max. $\Delta\varepsilon_m^{elast}$
Al matrix	69 – 50	23 – 32	σ_y [MPa]/$\Delta\varepsilon_m^{elast}$ [%]: ~40 / 0.06 (RT); ~20 / 0.04 (200°C)		
Si	165 – 157	2,5 – 4.0	20 – 28	0.22	<30
SiC	410-440	2.6 – 5.0	20 – 27	0.21	<30
Diamond (C_D)	Ca.1050	1 – 2	22 – 30	0.24	<27
Al_2O_3 Saffil® 3M-N610®	285 370	7 – 9	16 – 23	0.17	<38
C_f-HM ‖ long.	390	– 1	24 – 33	0.26	<25
⊥ radial	6	~10	10 – 20	0.15	<60
Mg matrix	45 – 35	25 – 35	σ_y[MPa]/$\Delta\varepsilon_m^{elast}$ [%]: ~50 / 0.1 (RT); ~10 / 0.03 (200°C)		
C_f-HM ‖ long.	390	– 1	26 – 36	0.31	<38
⊥ radial	6	~10	10 – 20	0.20	<100
C_f-HT ‖ long.	230	0-2	24 – 34	0.29	<42
⊥ radial	14	~10	10 – 20	0.20	<100

Table 2. MMC investigated and corresponding expansion data while heating from room temperature to about 100°C [18]

MMC type	Designation: matrix/reinforcement/ vol% type symbol	Processing [1]	ΔV_0 vol.mismatch [ppm/K]	CTE^{ROM} (Long./tr.) [ppm/K]	elast.*CTE* Long./tr. [ppm/K]	Exp.*CTE* Long./tr. [ppm/K]
PRM particle reinforcement	AlSi10Mg/SiC/10p	Stir cast	5	19	18	17
	AlSi7Mg/SiC/55p[C]	Centrif.cast	31	12	10	9
	Al/C_D/63p [19]	Gas pressure infiltrated preforms	42 /24*)	10	9 / 2*)	7
	Al99.5/SiC/70p AlSi7Mg/SiC/55p AlSi7Mg/SiC/70p		39 /17*) 26*) 17*)	13 12 9	10 / 5*), 8[7] 9.5*), 9[7] 7.5*), 7[7]	8 9.5 7
IMC*)	AlSi7Mg/R-SiC/85i		9	6.5	5.5, 5[7]	5
SFRM short fiber reinf.	AlSi1.1 /Al_2O_3/20s AlSi7-12/Al_2O_3/20s AlSi18/Al_2O_3/20s	Saffil™ preforms infiltr. squeeze casting	42 38 – 36 31	21 (7 / 49) 20 (7/45-43) 17 (7 / 38)	15 / 47 15 / 44-42 14 / 38	18 / 20 17 / 19-18 16 / 16
CFRM cont. unidirectional fiber reinforced	AlMg1/Al2O3-N610/65f-UD	Gas pressure infiltrated preforms of wound fibers	34	13 (7 / 16)	8.5 / 14.5	8 / 15
	Al/C-M40/70f-UD		38	7 (-1 / 18)	0.7 / 20	0 / 21
	$MgAl_{0.6}$/C-M40/$70f_{UD}$		41	7 (-1 / 19)	0.2 / 21	1 / 22
	Mg/C-HTA/60f-UD		35	10 (0 / 21)	3.1 / 12	2 / 20

Different reinforcement architectures have been selected for this study (see Table 2 and Fig.1). Fig.1a shows the particle reinforced Al-matrix (PRM) with monomodal SiC_p from LBI/France. The high volume fraction of 55 vol.% SiC was achieved along the outer surface of a centrifugal casting, because SiC is of higher mass density than the liquid matrix. As can be seen the particles in $AlSi7Mg/SiC/55p^C$ are embedded in the matrix, but occasionally touch each other. A densely packed distribution of diamond particles of about 350 μm in diameter has been achieved by gas pressure infiltration of a particle preform yielding $Al/C_D/63p$ [19]. Higher volume fractions of SiC are produced by infiltration of performs with trimodal particle size distributions [14] as shown for $AlSi7Mg/SiC/70_p$ in Fig.1b.

Fig.1c depicts an interpenetrating MMC (IMC) consisting of an open porous sponge produced by the incomplete sintering of SiC-particles (i) infiltrated by the AlSi7-matrix [14]. Al_2O_3-Saffil™ preforms with 20 vol% short fibers (s) distributed randomly planar were mechanically infiltrated by squeeze casting (SFRM) [15], the result of which is shown in orthogonal planes in Fig.1d and e. α-Al-dendrites (white) can be distinguished from eutectic AlSi12 regions (grey) within the matrix in Fig.1e. Continuous C- or Al_2O_3-fiber bundles (f) wound densely on a mandrel are used as preforms infiltrated by gas pressure to produce uni-directionally (UD) reinforced MMC samples (CFRM) [16]. Cross sections of these are shown in Fig.1f and g. The continuous fibers in the CFRM are still grouped in bundles separated by lines of reduced fiber packing [20]. The fibers are frequently touching each other within the bundles, particularly the highly not-wetting HM C-fibers in the Mg-matrix Fig.1f. In addition to the MMC shown, AlSi10Mg/SiC/10p produced by stir casting (by Duralcan) is considered as it contains SiC-particles completely embedded in the matrix [14].

The eutectic Si content of the Al-matrix is considered as an additional reinforcing constituent. The eutectically solidified Si forms bridges between the densely packed SiC in AlSi7Mg/SiC/70p [10, 14 and 21]. The short fibers in the AlSi12/Al2O3/20s are interconnected as well by Si-bridges formed in the AlSi12 matrix [15], which can be seen in Fig.1e. The investigated SFRM contain about 2 vol.% porosity from processing as identified by light microscopy and synchrotron tomography [22]. <1 vol.% porosity was determined by synchrotron tomography in AlSi7Mg/SiC/70p samples [21] and in MgAlo.6/C-M40/70f-UD samples along some of the fiber bundles [20].

The average volume mismatch ΔV_0 per 1K temperature change for the constituents of each MMC is calculated according to eq. (1) around 100°C and the results are given in Table 2. The volume expansion coefficient is taken to be $\alpha = 3\ CTE$ for isotropic constituents. The difference between the free expansion of the volume fraction v_p of discontinuous reinforcement embedded in the matrix is calculated using eq. (1a). The amount of v_p of the matrix is replaced by the less expanding ceramic reinforcement, which produces the volume difference ΔV_0.

$$\Delta V_0 = 3\ v_p\ \Delta CTE \quad (1a)$$

$$\Delta V_0 = v_f\,(\Delta CTE^{||} + 2\ \Delta CTE^{\perp}) \quad (1b)$$

$$\Delta V_0 = 3(1 - v_p)\Delta CTE \quad (1c)$$

$$\alpha_f = CTE^{||} + 2\ CTE^{\perp} \quad (1d)$$

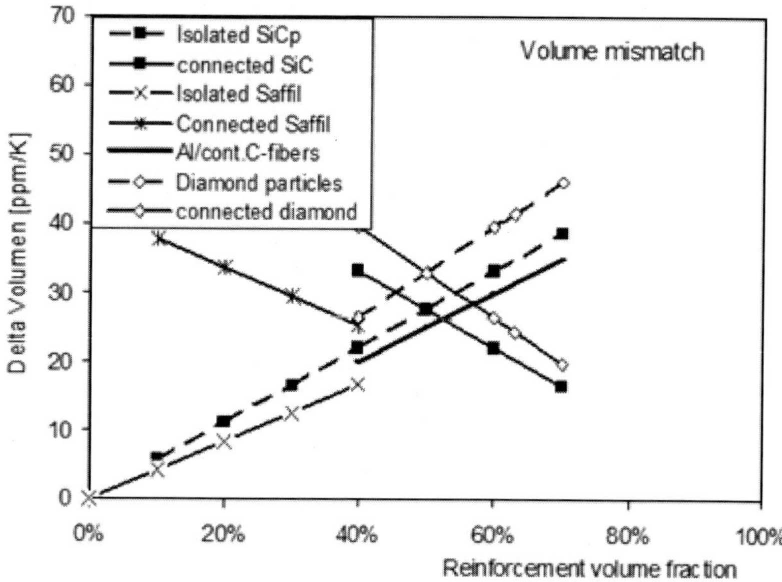

Figure 2. Volume mismatch per K assuming ROM for different volume fractions of reinforcements embedded in Al-matrix: discontinuous (SiC, diamond particles, or alumina short fibers) according to eq. (1a) compared with the same, but interconnected reinforce-ments according to eq.(1c); continuous unidirectional C-fibers according eq.(1b).

The parallel fibers in UD-CFRM allow the expansion of the matrix in between the fibers in radial direction. Therefore the fiber volume fraction v_f can be used to calculate ΔV_0 according to eq. (1b). The volume expansion coefficient of the anisotropic fibers α_f is calculated by eq.(1d). Thus the thermal mismatch volume is given by the expansion difference of the replaced matrix yielding eq. (1b). The matrix volume fraction $v_m = 1\text{-}v_p$ is inserted in eq.(1c), when the reinforcement is interconnected defining the boundaries for thermal expansion of the matrix. The volume mismatch in IMC is caused by the enclosed matrix tending to expand within the reinforcing sponge according to the CTE-mismatch. The volume mismatch increases with the volume fraction of discontinuous reinforcements embedded into the matrix, as shown in Fig.2. It decreases with the volume fraction of interconnected reinforcement, which results in a decreasing volume mismatch as the volume fraction of the reinforcement's sponge increases. The resulting values are presented in Table 2 for the different reinforcement architectures. The MMC processed by preform infiltration with Al-Si-melt, i.e. AlSi7Mg/SiC/70p, AlSi7Mg/SiC/55p and AlSi12/Al2O3/20s are considered as IMC owing to the contiguity between the eutectic Si with the reinforcement phase applying eq.(1c) as well as for AlSi7/Mg/R-SiC/85i.

2.2. Thermal Expansion of Composites

Dilatometry was applied to measure the change in length of samples of about 4x4x10-17 mm³ during heating and cooling at 3 K/min either from RT to 500°C [14,15] or for the CFRM from –50°C to 200°C [16] usually in consecutive cycles. The relative length change smoothed along intervals of 12K was differentiated with respect to temperature yielding an instantaneous linear coefficient of thermal expansion *CTE(T)* with an accuracy of ±1 ppm/K.

$$CTE_c^{ROM} = v_m CTE_m + (1 - v_m) CTE_r \tag{2}$$

$$CTE_{SFRM}^{ROM\perp} = 3\,CTE_c^{ROM} - 2\,CTE_r = (3v_r - 2)CTE_r + 3v_m CTE_m \tag{3}$$

$$CTE_{CFRM}^{ROM\perp} = v_r CTE_r^{\perp} + \tfrac{1}{2}(1 - v_r)(3.CTE_m - CTE_m^{||}) \tag{4}$$

The linear rule of mixture (ROM) eq. (2) refers to the thermal expansion of freely expanding constituents bonded in series. Thus it yields an upper limit of the *CTE* of composites excluding residual stresses. Data with suffix *m, r, c* refer to matrix, reinforcement and composite, respectively. If the matrix expands equal to the longitudinal fibers ($CTE_c^{||} = CTE_r^{||}$), it has to compensate the constraints in longitudinal direction by expansion transverse to the fibers to conserve its volume. The matrix in the unidirectional reinforced CFRM can expand in a cylinder symmetric way with the axis in fiber direction yielding eq.(4). The planar reinforced SFRM have to compensate the matrix expansion perpendicular to the fiber plane yielding a $CTE^{\perp}$ according to eq. (3).

Eq.(5) [8] yields upper and lower limits $CTE_c^{\pm}$, based on thermo-elastic behavior of the composite taking elastic straining of the constituents into account. E_m and E_r are the Young's moduli of matrix and reinforcement, respectively, where $E_r > E_m$. Examples, where $E_r < E_m$, are given in [23] resulting in a higher CTE of the composite than that resulting from ROM using eq.(2). The Poisson ratio of the considered reinforcement is taken to be 0.17. $K_c^{\pm}$ are the upper and lower Hashin-Shtrikman limits of the bulk module of the composite [24]. If K_m and CTE_m are larger than K_r and CTE_r [23], then the upper bound K_c^+ yields the lower limit CTE_c^- representing an interconnected reinforcement of low *CTE*, whereas the upper limit CTE_c^+ refers to particles with low *CTE* embedded in a high *CTE* matrix.

$$CTE_c^{+/-} = CTE_r + (CTE_m - CTE_r)\frac{1/K_c^{+/-} - 2/E_r}{1/E_m - 2/E_r} \tag{5}$$

$$CTE_c^{||} = \frac{E_r^{||} CTE_r^{||}(1 - v_m) + E_m CTE_m v_m}{E_r^{||}(1 - v_m) + E_m v_m} \tag{6}$$

$$CTE_c = \frac{K_r CTE_r (1 - v_m) + E_m CTE_m v_m}{K_r (1 - v_m) + E_m v_m} \tag{7}$$

$$CTE_{CFRM}^{\perp} = 1.3\,v_m CTE_m + 1.2\,v_r CTE_r^{||} - 0.25\,CTE_c^{||} \tag{8a}$$

$$CTE_{SFRM}^{\perp} = 2\left[CTE_{CFRM}^{\perp}\right] \tag{8b}$$

The upper limit of Schapery's eq.(5) is applied to MMC with particles embedded in the matrix. The lower limit yields the thermo-elastic result for reinforcements connected throughout the MMC. Thus, the lower limit of eq.(5) is applied for the calculation of the $CTE_c^{||}$ in fiber direction of the CFRM and in the fiber plane of the SFRM. Eq.(5) can be simplified to eq.(6) for the longitudinal $CTE^{||}$ of CFRM. E_m can be replaced by a strain hardening modulus of the matrix as soon as it plastifies [25]. In the case of ideal plasticity approximation, E_m becomes zero, which yields equal $CTE^{||}$ of the MMC and of the fibers in longitudinal direction, when inserted in eq. (6). Turner's model [7] can be applied to continuously reinforced composites using eq.(7), where the bulk modulus of the reinforcement can be replaced by $0.5E_r$ for the considered reinforcements. The transverse *CTE* of those MMC is calculated on the basis of volume conservation by inserting the corresponding *CTE*-values for CFRM in eq. (8a) and applying it to planar reinforced SFRM according to eq.(8b).

3. Results and Discussion

3.1 Particle Reinforced (PRM) and Interpenetrating Composites (IMC)

The experimental *CTE* at 100°C [14] for the MMC quoted as PRM and IMC are listed in Table 2. The isotropic *CTE* around 100°C resulting from ROM using eq.(2) are given there for comparison.

The corresponding data for the thermo-elastic *CTE* in Table 2 is obtained by using eq.(5) to calculate the upper limit of the composites' *CTE* for the stir cast PRM, the centrifugal cast PRM, the pure Al matrix composites Al/C_D/63p and Al99.5/SiC/70p, where the particles are assumed not to be bonded with each other. The elastic interaction of the constituents is effective as the *CTE(T)* calculated by ROM is significantly larger than the experimental results. The experimental *CTE(T)* for those PRM are slightly smaller than those calculated according to the upper Schapery limit, but larger than those calculated by the lower limit (marked by $^{*)}$).

The PRM produced by stir casting and centrifugal casting embed the particles almost completely into the ductile matrix. The maximum particle volume fraction for embedded SiC particles of monomodal size distribution was reached at 55 vol.%. The constraints to the thermal expansion of the matrix originate only from the interfaces of the particles surrounded by the matrix. As long as the expansion misfit can be compensated by elastic strains, the upper Schapery limit (eq.5) complies with this condition. The experimental *CTE(T)* of AlSi10Mg/SiC/10p and of the centrifugally cast AlSi7Mg/SiC/55p^C in Fig.3a are close to the upper Schapery limit calculated with the experimental *CTE(T)* of the Al-Si-matrix. The CTE at 100°C according to the upper Schapery limit is slightly higher than the experimental one (see Table 2).

The solubility of Si in Al increases from almost zero for T<300°C to 0.8 wt.% at 500°C. The drop in the *CTE(T)* curve of Al-Si-alloys above 350°C is due to the increasing dissolution of precipitated Si. This is because the atomic volume of Si precipitated in diamond structure

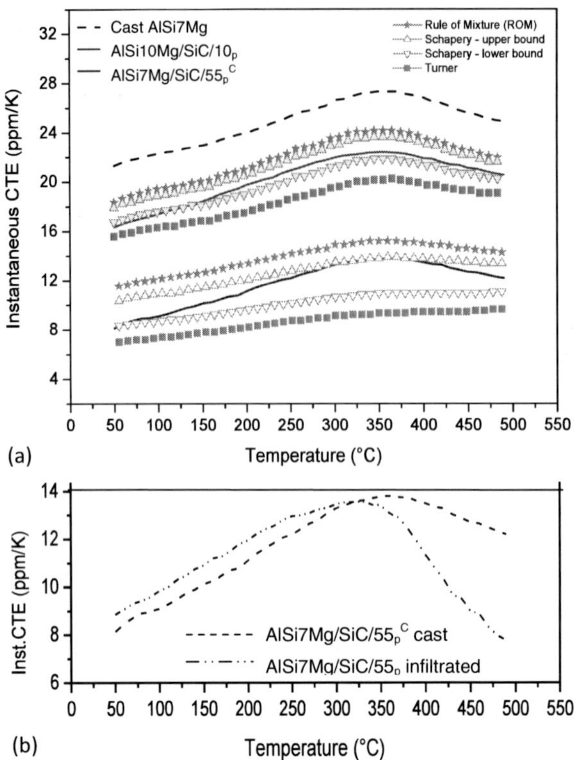

Figure 3. Comparison of CTE(T) of PRM: a) experimental results of ROM and of thermo-elastic models with the input of experimental CTE(T) for the matrix; b) experimental CTE(T) of centrifugal cast and infiltrated AlSi7Mg/SiC/55p.

is 20% larger than that in solid solution in Al [15]. The thermal expansion of both versions of AlSi7Mg/SiC/55p compared in Fig.3b is more or less the same up to 325°C complying with the thermo-elastic model. Above 325°C the *CTE(T)* of AlSi7Mg/SiC/55p, produced by melt infiltration of the monomodal SiC-powder preform, drops faster than the *CTE(T)* of the centrifugally cast sample.

Using ΔV_0 of Table 2, the volume difference of the centrifugal cast AlSi7Mg/SiC/55p^C version considered as PRM, where eq. 1a is applied, yields ΔV(500K)=1.55%. That of the infiltrated AlSi7Mg/SiC/55p version treated as IMC applying eq. 1c yields 1.3%. The difference in the measured volume expansion of the two AlSi7Mg/SiC/55p versions in that temperature interval of 500K amounts to <0.2 vol.%, which is of the same magnitude as the theoretical volume mismatch. At elevated temperature the yield strength of the matrix is decreasing and the thermally activated flow is enhanced. The SiC-particles in the preform touch each other and are frequently connected by the eutectic Si. The 3-dimensional network of low expanding reinforcement produces compressive stress onto the expanding matrix, which were verified by [26]. The plastically deforming matrix can penetrate pre-existing voids. Thus a void fraction about 0.2 vol.% is enough to produce the observed difference in *CTE(T)*. It can be assumed that even a high quality gas pressure infiltration leaves a porosity of that magnitude in the product owing to the higher shrinkage of the matrix than

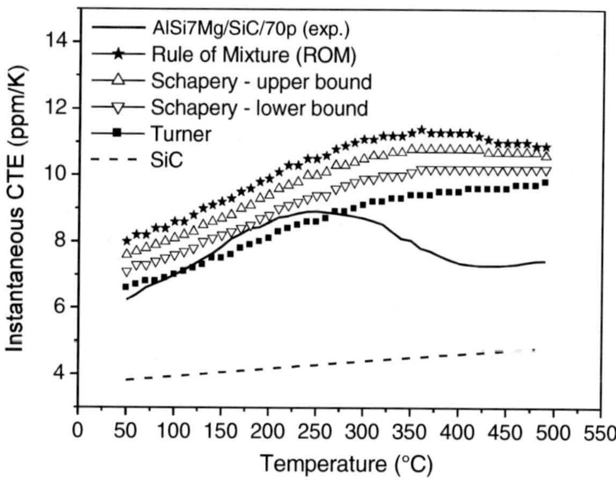

Figure 4. Experimental *CTE(T)* of AlSi7Mg/SiC/70p during heating deviating from the results of the thermo-elastic models >250°C; the *CTE(T)* of SiC is presented for comparison.

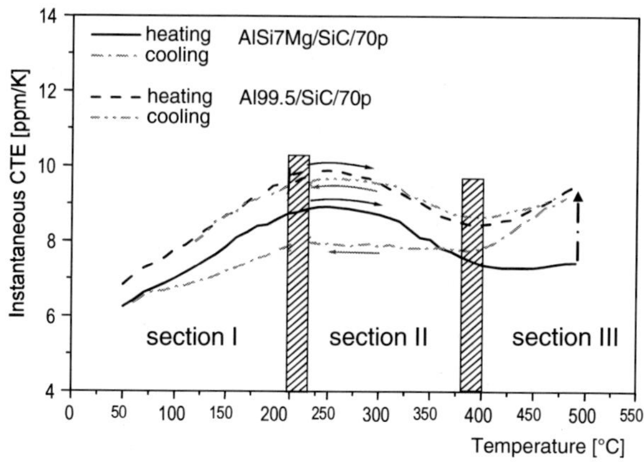

Figure 5. Comparison of *CTE(T)*of AlSi7Mg/SiC/70p and Al/SiC/70p measured during heating and cooling (note the jump in the CTE of AlSi7Mg/SiC/70p at the temperature inversion at 500°C); The curves are divided into 3 sections (see text).

agglomerations of packed particles. The lower bound of Schapery's eq. (5) is used for the gas pressure infiltrated composites with AlSi7-matrix: AlSi7Mg/SiC/55p, AlSi7Mg/SiC/70p, and AlSi7/Mg/R-SiC/85i, which are considered as IMC.

The results of eq. 7 are indicated as well in brackets for some AlSiC materials in Table 2. The results according to Turner's eq. (7) are slightly below those of the lower Schapery limit. The experimental results of those MMC, quoted above as IMC, correspond well to the Turner model, the results of which are very close to the lower Schapery limit (see Fig.4, Table 2).

The presence of the eutectic Si (ca.1.6 vol% of the MMC) in AlSi7Mg/SiC/70p reduces the *CTE(T)* by ~1 ppm/K in the whole temperature range with respect to the *CTE(T)* of Al/SiC/70p (see Fig.4). The *CTE(T)* curves of the two matrices indicate similar thermo-elastic behavior in this temperature region. Assuming that the SiC (or the diamond particles) are compressed during infiltration, the preform consists of touching, partly interlocked particles without Si-bridges, which can hardly be moved by the expanding matrix during heating. The experimental *CTE(T)* results are slightly higher than those according to the IMC interpretation in Table 2. Thus, it can be assumed that the condition of completely embedded particles is not fulfilled. The *CTE(T)* curves in Fig.5 are divided into 3 sections.

Section I represents the thermo-elastic range in heating and cooling, where the thermo-elastic models fit. In section II the matrix stress becomes compressive during heating, which makes the matrix flow into pre-existing voids [26]. Therefore the porosity decreases and consequently the total volume of the sample expands less than thermally expected. In section III, as soon as most of the voids are closed, the *CTE(T)* remains constant for the AlSi7Mg-matrix and increases slightly for the pure Al-matrix, where elastic stresses build up again. The hysteresis of the *CTE(T)* for AlSi7Mg/SiC/70p, when changing from heating to cooling in Fig.5, is explained by the relaxation of higher elastic strains in the interconnected reinforcement than in Al/SiC/70p without Si-bridges [14].

The volume mismatch for the whole temperature amplitude amounts to 0.9 vol.% for the pure Al-matrix and 0.8 vol.% for the alloy matrix (see Table 2). Such a mismatch can only be partly compensated by an elastic strain amplitude (0.1-0.2 vol.%). +0.1% strain in the matrix at RT corresponds to 70 MPa tensile stress and -0.1% at 500°C to 50 MPa compressive stress. A 0.2 vol.% difference remains between the experimentally determined expansion and that corresponding to an hypothetic thermo-elastic behavior in agreement with the change in porosity measured by synchrotron tomography [26]. The *CTE(T)* of Al/SiC/70p for cooling coincides with the heating *CTE(T),* which implies that the voids closed during heating are reversibly opened again during cooling causing negligible hysteresis.

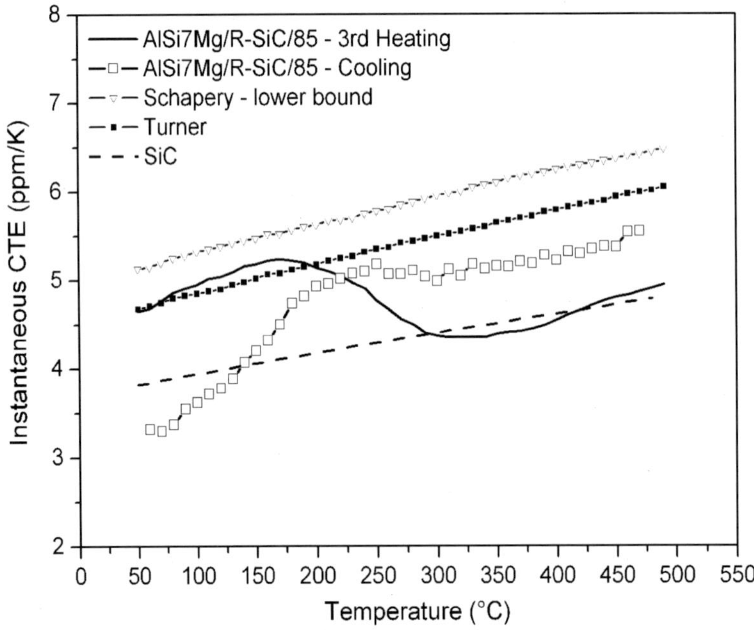

Figure 6. Experimental *CTE(T)* of the IMC AlSi7Mg/R-SiC/85i during heating and cooling compared with the thermo-elastic models and the *CTE(T)* of SiC.

The Al/C_D/63p behaves similarly. The *CTE(T)* of AlSi7Mg/SiC/70p during cooling is different from that during heating. The *CTE(T)* of AlSi7Mg/SiC/70p in cooling starts at 500°C by about 2 ppm/K higher than during heating. This effect is correlated to the relaxation of elastic strains in the Si-bridges connecting the SiC particles in section III producing a small spring back effect as soon as the cooling starts. The *CTE(T)* remains constant in section II, where compressive stresses in the matrix invert into tensile stresses, which open some voids again counteracting the decrease of the *CTE(T)* of the constituents.

The rigid connectivity of the sintered R-SiC preforms produces a *CTE(T)* as shown in Fig.6. The elastic expansion up to about 200°C agrees with the thermo-elastic model according to Turner (see Table 2). The *CTE(T)* drops to that of the reinforcing SiC with increasing temperature. The volume mismatch of 0.4 vol.% is compensated completely by filling pre-existing voids during heating. The experimental *CTE(T)* is about 0.5 ppm/K below the thermo-elastic level during cooling down to about 200°C indicating some elastic straining followed by re-opening of voids. During further cooling the *CTE(T)* drops to that of SiC and even below due to elastic contraction of the SiC structure by the shrinking matrix. There is a significant thermal hysteresis in the heating and cooling *CTE(T)* due to the connectivity of the reinforcement.

3.2 Short Fiber Reinforced Composites (SFRM)

Fig.7a shows the anisotropy of the percent length change (*PLC*) for the different Si-content of the investigated AlSiX/Al2O3/20s [15]. The $PLC(T)^{||}$ refers to the random fiber plane and exhibits only a negligible increase with increasing Si-content yielding a $CTE(T)^{||}$ essentially independent from the matrix composition. The *CTE* calculated by applying eq.(2) corresponds well to the measured *CTE* perpendicular to the fiber plane in the elastic range up to about 100°C (see Table 2). The in plane *CTE* and the *CTE* perpendicular to the fiber plane of the SFRM are calculated using eq. (3) as given in brackets in Table 2. Those predictions of the *CTE* at 100°C by the ROM do not yield the experimental values nor do the thermo-elastic models (see Fig.8). Above 100°C, the $PLC(T)^{\perp}$ perpendicular to the fiber plane becomes

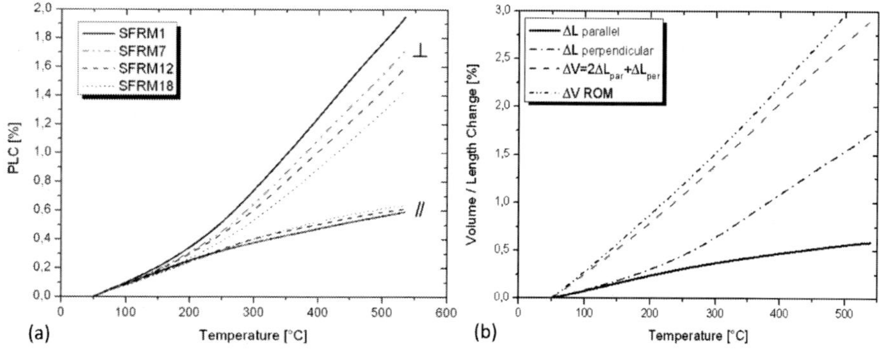

Figure 7. Thermal expansion of SFRM [15]: a) percent length change (*PCL*) of AlSi1-18/Al2O3/20s in fiber plane || and perpendicular ⊥, b) *PCL(T)* of AlSi7/Al2O3/20s in both directions and the resulting volume expansion compared with ROM results.

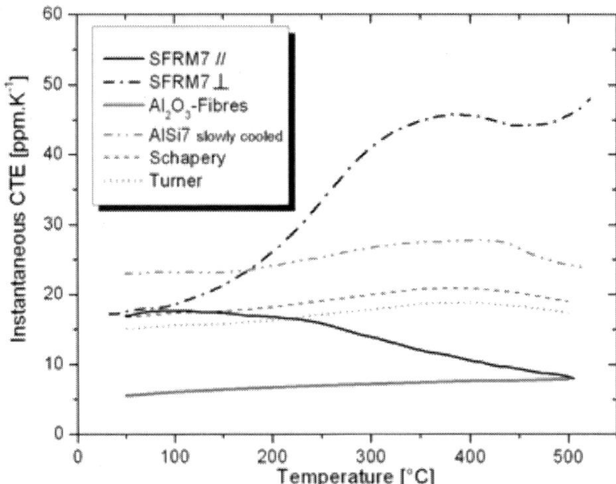

Figure 8. *CTE(T)* of AlSi7/Al2O3/20s in fiber plane and perpendicular during heating compared with results of isotropic lower Schapery limit and of Turner model; both are based on the experimentally determined *CTE(T)* of the matrix and that of the Al2O3 reinforcement as depicted.

much larger than in fiber plane and decreases with increasing Si-content as the volume mismatch is decreasing (see Table 2). Fig.7b shows the experimental *PLC(T)* in fiber plane and perpendicular together with the calculated volume expansion. According to the rule of mixture the volume expansion is 0.3 vol.% larger than the experimental result up to 500°C. This cannot be compensated for by elastic strains, but by reducing the void volume fraction by the plastifying matrix. The *CTE(T)* curves for SFRM7 in Fig.8 show the continuous decrease of the in plane $CTE(T)^{\|}$ from the thermo-elastic lower Schapery limit above 150°C to the *CTE* of the fibers at 500°C. The *CTE(T)* of the matrix is depicted as well, which is used for the calculation for the *CTE(T)* according to eq.(3). The matrix gets compressed in the fiber plane expanding significantly more perpendicular to the fiber plane.

$CTE(T)^{\perp}$ increases to 45 ppm/K reaching 350°C and remains at this level up to 500°C. A small reduction in expansion is attributed to the dissolution of precipitated Si, but most of it (~0.2 vol.%) is produced by filling of pre-existing voids in the SFRM samples. The transverse connectivity increases with increasing Si content [15], which accelerates the filling of voids during heating. The application of the lower Schapery limit eq. (6) for continuously reinforced composites to SFRM seems justified as long as the SFRM expand more or less isotropic from RT to 100°C owing to the interconnected 3-dimensional network of short fibers with the eutectic Si bridges. The anisotropic ROM predictions in Table 2 correspond rather to the elevated temperature results, where elastic stresses become negligible due to the matrix expanding into voids (see >400°C in Fig.8).

3.3. Unidirectional Fiber Reinforced Composites (CFRM)

Neither the longitudinal *CTE* of CFRM inserting $CTE_r^{\|}$ into eq.(2) nor the CTE in both directions for reinforcements by anisotropic fibers using eq.(4) can be predicted by application of ROM (see Table 2). The predictions of $CTE_c^{\|}$ and $CTE_c^{\perp}$ for CFRM with isotropic alumina fibers fit reasonably well to the experimental results. The predictions of the thermo-elastic models agree well with the *CTE* measured in both directions within the elastic range of the matrix. The dimensional changes in the fiber direction are limited by the elastic deformation of the fibers. Unidirectional (UD) orientations of continuous alumina fibers reduce the longitudinal thermal expansion much more than randomly oriented alumina short fibers do. Fig. 9a shows the hysteresis of UD alumina fiber reinforced Al during 2 thermal cycles from -50°C to 120°C, followed by 2 cycles up to 200°C. The dilatometer measurements are evaluated from -45°C to 195°C to allow for smoothing of the curves within 6 K from the inversion temperatures.

The thermal expansion of AlMg1/Al2O3-N610/70f-UD in fiber direction is almost equal to the expansion of the free fibers (see Tables 1 and 2). There is almost no hysteresis for the cycles to 120°C, but it opens up for the 200°C cycles indicating plastic deformation of the matrix at elevated temperature, where the thermo-elastic models fail. The dominance of the fiber is evident for Al/C-M40/70f-UD in Fig.9a: the CFRM expands by 0.008% up to 120°C, and no more up to 200°C, whereas the free fiber would shrink by -0.016% and -0.024%, respectively. The difference is made up by elastic straining of fibers and matrix, which exerts at least 10-20 MPa tension on the fibers at elevated temperature. The hysteresis of the *PLC*

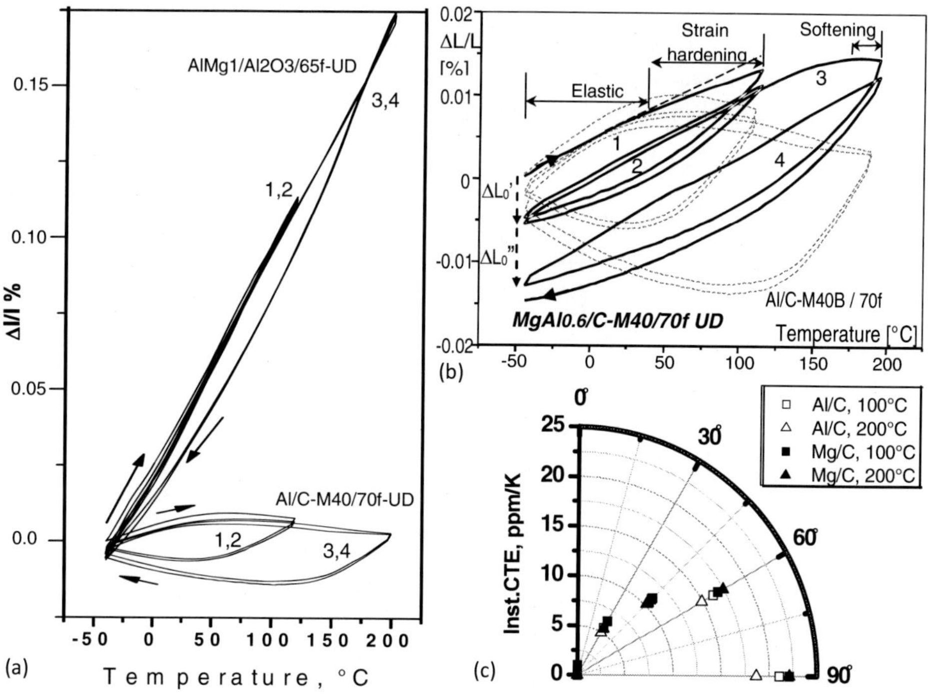

Figure 9. Linear expansion of unidirectional CFRM measured during cycling twice between –50°C and 120°C, then –50°C and 200°C (cycles 1-4): a) continuous HM-C-fiber and alumina fiber reinforced Al-matrices; b) comparison of HM-C-fiber reinforced pure Al and MgAl0.6 (with indications of the matrix deformation mode). ΔL0',” indicate the macroscopic shrinkage of the MgAl0.6/C-M40/70f-UD sample after changing Tmax of the thermal cycle. c) Anisotropy of the CTE of the same CFRM as in (b) determined at 100°C and 200°C.

during both cycling amplitudes indicates plastic deformation. Fig.9b illustrates the matrix effect by comparing Al- and Mg-matrix reinforced with the same type of fibers. The enlarged scale allows us to relate the deformation mechanisms for MgAl0.6/C-M40/70f-UD.

The elastic straining during the first heating ends at about 40°C, where plastic deformation with some strain hardening (probably by twinning) of the matrix starts. When cooling begins, the compressive elastic strains of the matrix relax first, then invert into tensile stresses during cooling below 40°C compressing the expanding fibers in longitudinal direction. This produces a longitudinal contraction of the CFRM by ΔL_0'. The second heating to 120°C causes mainly elastic interactions due to the hardening produced during the first cycle. There is no length change of the CFRM after the 2nd cooling. The next heating up to 200°C does not produce any elongation during the last 20°C indicating that the elastic stresses relax and hardening partly recovers. During cooling from 200°C stress inversion occurs at a higher temperature than in the previous cycle. The tensile stresses in the matrix increase while cooling; thus the fibers are compressed more than before producing a further longitudinal contraction of the CFRM by ΔL_0". During the following heating the purely elastic interaction lasts until about 50°C followed by some hardening until 200°C. Cooling reproduces the stress situation as in the previous cooling period. Leaving the sample for about one day at room temperature causes relaxation of the matrix and the thermal cycling behavior starts again as shown [16]. The comparison with the *PLC(T)* of Al/C-M40/70f in Fig.9b shows that the elastic range for the Al-matrix is smaller (<RT), and the strain hardening starts to recover at about 50°C. The residual stresses after cooling are much smaller in the soft Al-matrix than in the Mg-matrix yielding very little longitudinal contraction (<ΔL_0') of the CFRM sample after thermal cycles.

Unidirectional CFRM exhibit a more pronounced anisotropy in thermal expansion than SFRM. The results for the instantaneous *CTE* of Al and MgAl0.6/C-M40/70f-UD at 100 and 200°C are plotted in Fig. 9c for different directions with respect to the fiber orientation. The *CTE* of the CFRM with the Al-matrix is little smaller than that with the Mg-matrix. This is, because the yield strength of the Mg-matrix is higher than that of the pure Al-matrix in the considered temperature range. In both cases the small longitudinal expansion is compensated by high radial $CTE_c^{\perp}$. The thermo-elastic calculations using eq. (6) yield longitudinal *CTE*-values very close to the experimental results (see Table 2). The transverse *CTE* according eq. (8a) fits only for the isotropic alumina fibers. Simple isotropic calculations by ROM according eq. (2) are not useful for anisotropic C-fiber reinforcements.

The volume expansion of UD-CFRM calculated by $\Delta V(T) = \Delta L^{||}(T) + 2\ \Delta L^{\perp}(T)$ is depicted in Fig. 10a for Al/C-M40/70f together with the free volume expansion of the constituents composed by ROM. The heating-cooling hysteresis between the temperatures -45°C and 195°C opens to about 0.1 vol.%. The differentiation of the curves in Fig.10a) yields the volumetric expansion coefficient α depicted in Fig.10b. The $\alpha(T)$ curves show a clear drop above 70°C during heating and at T<100°C during cooling. These effects are interpreted as closing of the pores by about 0.1 vol.% during heating and re-opening during cooling by the plastifying matrix. Thus it is assumed, that pores between touching fibers can change their volume by expansion and contraction of the matrix as it was shown by insitu-synchrotron tomography during thermal cycling [20].

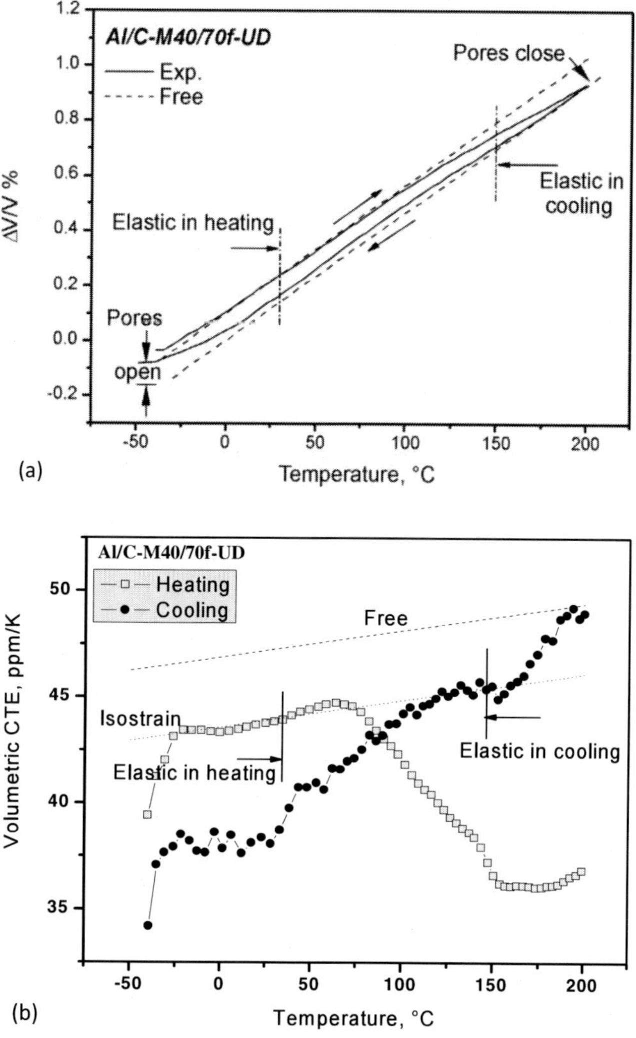

Figure 10. Volumetric thermal expansion of Al/C-M40/70f-UD between -45°C and 195°C: a) Relative volume change during heating and cooling calculated from the measured linear expansions $CTE_c^{||}$ + $2.CTE_c^{\perp}$ compared with the volume changes using ROM for α according to eq.(2) and (4). The elastic regions are marked, whereas the pore closing and opening by plastic deformation cause the hysteresis; b) corresponding volumetric expansion coefficient α in the elastic ranges for about 100K during heating and to about 150°C during cooling compared with the isostrain ROM calculation; the decrease of α by pore closing during heating and reopening during cooling.

The limits of the linear, longitudinal $CTE_c^{||}$ can be calculated using eq.(6) by inserting either the Young's modulus of the matrix or the strain hardening modulus, which can converge to zero assuming ideal plasticity [25]. The experimental results for 2 thermal cycles are plotted in Fig.11, where the CTE-levels for purely elastic and ideally plastic matrix deformation are indicated. From -40°C to about 0°C the matrix deforms elastically. Then plastification of the matrix begins due to compressive stresses imposed by the fibers represented by a decreasing strain hardening modulus with increasing temperature. The

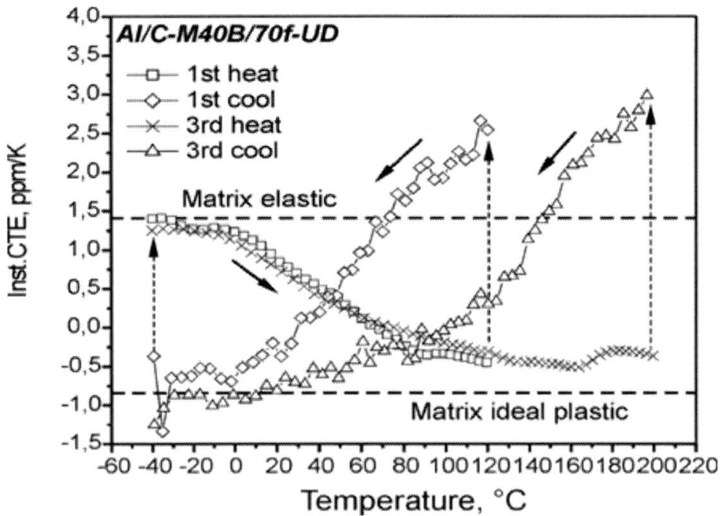

Figure 11. Longitudinal $CTE_c^{\parallel}(T)$ of Al/C-M40/70f-UD determined by thermal cycles: 1st from –50°C to 120°C and a 3rd cycle from -50°C to 200°C as indicated by the arrows. The increase of the $CTE_c^{\parallel}$ when changing from heating to cooling is marked. The modelling limits according to eq.(6) for elastic and ideal plastic matrix are indicated.

matrix behaves as an ideal plastic in a small strain range above 80°C. When cooling starts, the $CTE_c^{\parallel}$ jumps to values well above the elastic model independent of the temperature of inversion. As the thermal expansion of the matrix is inverted to contraction, the elastic tension of the fibers is released producing a spring back effect on the composite. That causes an increase of the apparent $CTE_c^{\parallel}$ for longitudinal contraction of the CFRM at the beginning of the cooling period.

Then the shrinking matrix is plastifying in tension with reducing strain hardening modulus until reaching ideal plasticity. This is reached at higher temperatures the higher the temperature of inversion of the thermal cycle, i.e. at RT when cooling from 120°C and at ca.70°C cooling from 200°C, thus requiring more or less a similar temperature interval for the transition from elastic to practically ideal plastic matrix deformation of pure Al-matrix.

Conclusion

Dilatometry is a sensitive tool to reveal the transition from thermo-elastic to thermo-plastic or viscous matrix deformation in MMC during thermal cycling. The derived *CTE(T)* of really discontinuously reinforced metals can be predicted by the upper Schapery limit. Actually all MMC produced by gas pressure infiltration of preforms are to be considered as interpenetrating MMC, i.e. IMC: the high volume fractions of particles, which cannot be compressed; particularly volume fractions of particles >55 vol.% or short fiber reinforcements interconnected by Si from the Al-Si-matrix. The unidirectional continuous fiber reinforced CFRM behave like those in fiber direction as the matrix expansion is forced into the transverse directions. The range of thermo-elastic behavior can be calculated by the lower Schapery limit or the Turner model. ROM allows us to estimate the elevated temperature limit of *CTE(T),* where elastic stresses become negligible. The transition range in between can be estimated using the strain hardening moduli approaching zero for the matrix instead of the Young's moduli.

The instantaneous *CTE(T)* of fiber reinforced MMC present anomalies caused by anisotropic plastification of the matrix. Dilatation measurements in the plane of random fiber orientation of SFRM or along the fibers of UD-CFRM together with measurements in transverse directions reveal the anisotropy of their thermal expansion behavior. Matrix plastification increases the thermal expansion transverse to the fiber orientation.

All MMC with densely packed or interconnected reinforcements (IMC) contain pores from processing and from matrix shrinkage occurring during cooling from the processing temperature. The porosity can be estimated by the volume misfit between the constituents using the appropriate eq.(1). The calculation of the volume expansion from linear measurements enables the detection of differences with respect to volume conservation, when compared with the thermal expansion of the constituents taking into account elastic strain accommodations. Small temperature dependent variations of the pore volume fraction in MMC can be identified, which are characteristic for IMC. These porosities <1 or 2 vol.% seem to reduce damage sensitivity during thermal cycling by avoiding instantaneous debonding.

ACKNOWLEDGMENTS

We are grateful to L.Weber, EPF-Lausanne (CH) for helpful discussions improving the content and to A.Rhyswilliams for language corrections.

REFERENCES

[1] Clyne T.W. "*Metal matrix composites*", Comprehensive composite materials, ed.by Kelly A. and Zweben C., vol.3, Elsevier, Oxford 2000.

[2] Zweben C., *J.Metal*, 44 (1992)15-23.

[3] Schellin P.K., Shi L., Goodson E., *mat.today*, 6 (2005) 30-35.

[4] Zweben C., *J.Adv.Mater.*39 (2007) 3.

[5] Evans A., San Marchi C., Mortensen A. "*MMC in Industry*", Kluwer Ac.Publ., Dordrecht 2003.

[6] Delannay F., "*Thermal stresses and thermal expansion in MMC*" in [1], 341-369.

[7] Turner P.S., *J.Re.Nat.Bu.Stand.*, 37 (1946) 239-250.

[8] Schapery R.A., *J.Comp.Mater.*, 2/3 (1968) 380-404.

[9] Shen Y.L., Needleman A., Suresh S., *Met.Mat.Trans.* 25A (1994) 839-850.

[10] Nam T.H., Requena G., Degischer H.P., *Comp.Sci.*A 39 (2008) 856-865.

[11] Roudini G., Tavanger R., Weber L., Mortensen A. *Int.J.Mat.Res.*101/9 (2010) 1113-1120.

[12] Shen Y.L., *Mat.Sci.Eng.*A237 (1997) 102-108.

[13] Balch D.K., et al., *Met.Mat.Trans* 27A (1996) pp.3700.

[14] Huber T., et al., *Comp.Sci.Techn.* 66 (2006), 2206-2217,.

[15] Lasagni F., et al., *Kovove Mater*, 44 (2006) 65-73,.

[16] Aly M., PhD thesis, Vienna Univ. of Technol., Vienna 2004.
[17] Ehrenstein G.W., *"Faserverbund-Kunststoffe"*, Hanser Verlag, München 2006.
[18] Degischer H.P., Lasagni F., Schöbel M., Huber T., Aly M., *Proc. "Int.Conf.on Advanced Materials and Composites* - ICAMC 2007, www.niist.res.in/icamc2007/proceedings.htm, Paper-Nr. I-1, 11p.
[19] Weber L., Tavanger R., *Adv.Mat.Res.* 59 (2009) 111-115.
[20] Rodriguez Hortalá M., Requena G., Seiser B., Degischer H.P., Di Michiel M., Buslaps T., *"3D-Characterization of continuous reinforced composites";* Proc.17th Int.Conf.Composite Materials-ICCM17, Edinburgh/Scotland, IoM (2009), 9 p.
[21] Schöbel, M., Requena, G., Degischer, H.P., Kaminski, H., Buslaps, T., Di Michiel, M., *J.Mat.Sci.& Eng. Technol.*, 38/11 (2007), 927-933.
[22] Requena G.C., Degischer H.P., Marks E.D., Boller E., *Mat.Sci.Eng.* A 487/1-2 (2008), 99-107
[23] Vetterli M., Tavanger R., Weber L., Kelly A., *Scr.Mat.*64 (2010) 153-156.
[24] Hashin Z., Shtrikman S., *J.Mech. & Physics of Solids* 11 (1963) 127.
[25] Böhm H.J., et al., *Comp.Eng.*, 5 (1995) 37-49.
[26] Schöbel M., Fiedler G., Degischer H.P., Altendorfer W., Vaucher S., *Adv.Mat.Res.* 59 (2009) 177-181.

In: Metal Matrix Composites
Editor: J. Paulo Davim

ISBN: 978-1-61209-771-8

Chapter 4

OPTIMIZATION OF PROCESS PARAMETERS DURING STIR CASTING OF AL/5VOL% GR_P AND AL/5VOL% AL_2O_3-MMC

Alakesh Manna*
Department of Mechanical Engineering, PEC University of Technology; (Formally Punjab Engineering College); Chandigarh; India

ABSTRACT

This paper presents an experimental investigation into the influence of stir casting parameters during casting of Al/5vol%Gr_p and Al/5vol%Al_2O_3-MMC. Taguchi method based design of experiment is used to optimize the stir casting process parameters for effective production of Al/Gr_p and Al/Al_2O_3-MMC,s. According to the Taguchi quality design concept an L_{27} (3^{13}) orthogonal array has been used to determine the S/N ratio (dB), ANOVA and 'F' test values for indicating the most significant parameters affecting the stir casting performance characteristics such as micro hardness and tensile strength of the cast metal matrix composites samples. Test results indicated that the stirring speed and stirring time are the most significant and significant casting parameters for micro hardness of Al/5vol%Al_2O_3-MMC. The interaction between pouring temperature and stirring speed is the most significant parameter for tensile strength of Al/Gr_p and Al/Al_2O_3-MMC,s. Utilizing experimental results and using Gauss elimination method mathematical models relating to the micro hardness and tensile strength are established to investigate the influence of casting parameters during stir casting of MMC,s.

1. INTRODUCTION

Metal matrix composites have different important properties such as high specific strength and stiffness at elevated temperatures and good creep, fatigue and wear resistance.

* Corresponding author, E-mail: kgpmanna@rediffmail.com

These properties are not achievable with lightweight monolithic metal or alloys individually. Particulate metal matrix composites have nearly isotropic properties when compared to long fiber reinforced composite. Aluminum-silicon alloys(Al-6063) as a matrix material, are characterized by light weight, good strength-to-weight ratio, ease of fabrication at reasonable cost, good thermal conductivity, excellent corrosion and wear resistance properties. Amongst various processing routes stir casting is one of the promising routes available for fabrication of composite. The process is simple, flexible and applicable for large quantity production. The cost of producing the MMC's by stir casting is 1/3rd to half of other competitive method. In the present experimental work, the basic aim is to identify the effect of stir casting parameter for better mechanical properties of stir cast MMC,s. The Taguchi method based robust design is used to formulate experimental layout, to analyze the effect of stir casting parameters namely pouring temperature, stirring time, and stirring speed on the mechanical properties such as micro hardness and tensile strength, and to predict the optimal setting for stir casting process parameter.

Seo and Kang (1999) fabricated the metal matrix composites by melt stirring. Author used the prepared composite specimens for studying the mechanical properties and compared the results with the extruded Al-specimens. Manoharan, and Gupta (1999) fabricated metal matrix composites by stir casting route and cast the mixed Al-and particles by pouring into the prepared mould. Authors concluded that work hardening behavior of composites with the different SiC volume fraction can be modeled using a modified continuum theory. Hashim, et al (1999) studied in detail the stir casting process and concluded that the stir casting is generally accepted as a particularly promising route as it is simple, flexible and has applicability to large quantity production. W.Zhou and Z.M Xu (1997) prepared two types of SiC particulate reinforced composites by gravity casting and concluded that the SiC particles were observed to be located predominantly in interdendritic region and act as substrates for heterogeneous nucleation of Si crystals. Naher, et al (2003) simulation studied on stir casting process where liquid and semisolid aluminum were replaced by other fluids. Authors concluded that change in viscosity had tremendous effect on SiC dispersion and settling time. McKimpson, et al (1993) studied on the characteristics properties of commercially available discontinuously reinforced aluminium metal matrix composites through multi-laboratory testing program and concluded that these materials could be selected reliably for airframe structural application where its high strength and good damage tolerance are advantageous. Hunt, et al (1993) studied on the microstructure and the microstructural factors for various fracture modes for detection of fracture resistance of discontinuous reinforced aluminium metal matrix composite. They concluded that the thermo-mechanical properties could be improved through spatial distribution of reinforcement particle during thermo- mechanical processing of SiC- reinforced/ Al-metal matrix composite. Ourdjini et al (2001) studied on settling of particulate matter during casting of metal matrix composites. Authors used A356 Al-Si matrix with 7 wt % and 20wt % SiC reinforced particles and concluded that settling measurements during isothermal holding showed that SiC particles settled at much slower rate than predicted from theoretical model. Hashim et.al. (2002) studied the effect of parameters on the final distribution of the particles in the matrix alloy and concluded that in order to have a homogeneous distribution of reinforcement, factors like particle density, size, shape, volume fraction and surface properties have considerable affect on mechanical properties. Nai and Gupta (2003) studied the effects of different stirrer geometries (two

Table 1. Chemical composition of Matrix Alloy

Element	Cu	Sn	Pb	Si	Mg	Zn	Mn	Sb	Fe	Ti	S	Al
%	0.01	-	.01	1.48	0.02	0.001	0.01	-	0.21	0.01	-	Rest

Table 2. Casting parameters and their level

SNo	Casting parameters	Level		
		1	2	3
1	A :Pouring Temperature, ^{0}C	750	825	900
2	B : Stirring speed, rpm	150	225	300
3	C : Stirring time, min	3	6	9

bladed, four bladed and circular shaped) on the synthesis of Al/SiCp-particles functionally gradient materials. Authors concluded that the two bladed stirrers yield most satisfactory result as regards to uniform distribution, porosity levels and micro hardness when compared to other stirrer geometries.

2. Planning for Experimentation

In the study, aluminium ingot was cleaned so as to remove surface impurities. Cleaned ingots were melted in an electric furnace. The temperature was raised 50^{0}C above liquidus temperature. Alumina and graphite particulate were preheated to 950^0 C for 90 minutes to enhance their wettability with aluminum matrix. The molten aluminum was then cooled so as to attain a semi solid stage. At this point alumina and graphite particulate were added and the mixer stirred at 220 rpm on a developed stir casting set up. The slurry was again reheated to 50^0 C above liquidus temperature and poured in a preheated permanent mould. The chemical composition of matrix is given in table 1.

The different sets of experiments have been carried out during stir casting operation utilizing three furnaces simultaneously. Micro hardness of the prepared casing samples have been measured by HMV micro hardness tester of capacity range 10 - 1000g at different positions along the prepared surface of the cast samples. The tensile strength of the prepared casting sample has been measured utilizing Universal testing machine HT-2107A (Hungta, having 310 tons capacity). The average test results are taken for optimization of the casting parameter. According to the Taguchi method based robust design suggested by Montgomery, (1997); an L_{27} (3^{13}) orthogonal array is employed for the experimentation. Three casting parameters such as pouring temperature, stirring speed, and stirring time are considered as controlling factors and each parameter has three levels, namely small, medium and large are denoted by 1, 2 and 3 respectively. Table 2 represents the casting parameters and their levels considered for experimentation.

The first column, second column and sixth column of L_{27} (3^{13}) array are assigned to the pouring temperature (A), stirring speed (B) and stirring time (C) respectively. Initially, 27 sets of experiments have been performed according to the L_{27} (3^{13}) orthogonal array, each set of experiment repeated three times and total 81 experiments have been conducted for the present investigation.

3. Experimental Results and Discussion

Figure 1 shows the microstructure of prepared Al/5%vol Al_2O_3-MMC at 100X, this sample was prepared at 825^0C pouring temperature, 225 rpm stirring speed with 6 min continuous stirring time. Figure 2 shows the microstructure of prepared Al/5%vol Gr_p-MMC at 100X, this sample was prepared at 750^0C pouring temperature, 150 rpm stirring speed with 6 min continuous stirring time.

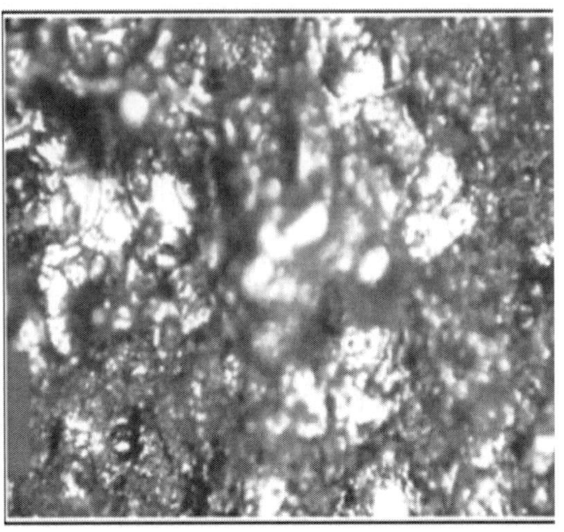

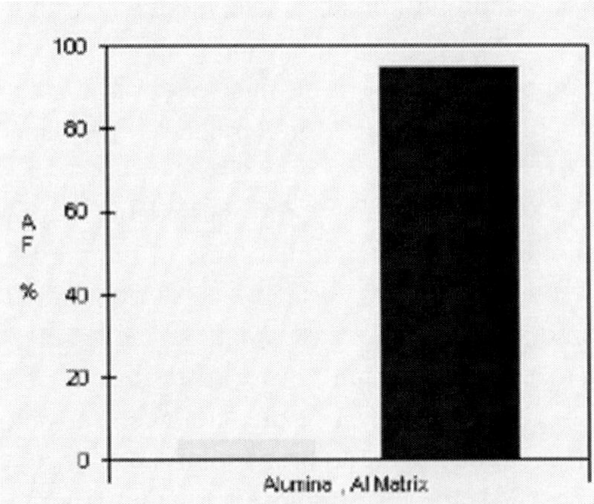

SID:Alumina 5%	Frame area(mm^2)	Alumina(%)	Al Matrix(%)
Frame-1	0.2272	4.975	95.025
Average	0.2272	4.975	95.025

Figure 1. Microstructure of prepared Al/5%vol Al_2O_3- MMC at 100X.

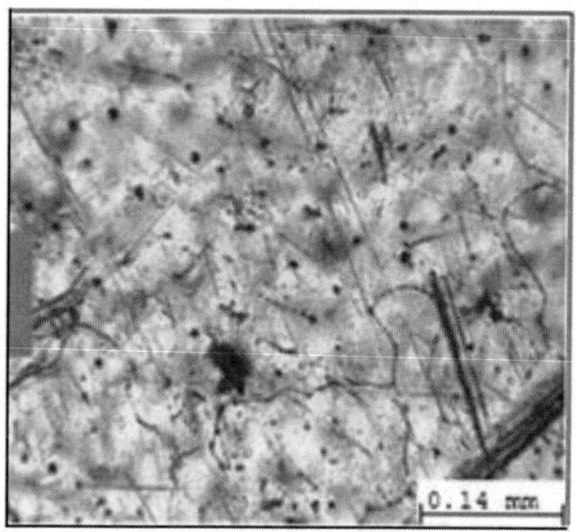

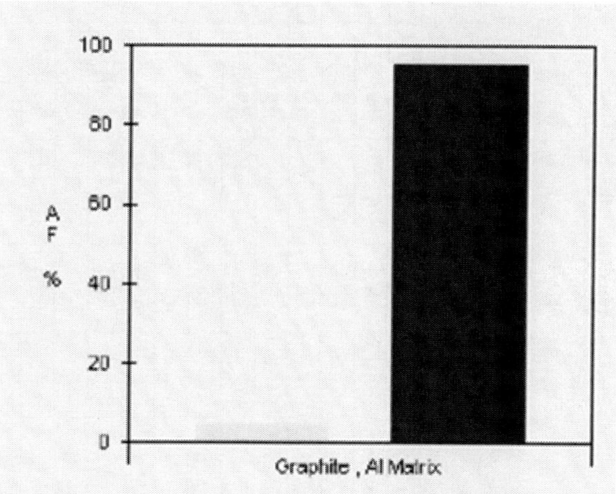

SID:Graphite 5%	Frame area(mm^2)	Graphite (%)	Al Matrix(%)
Frame-1	0.2272	4.671	95.329
Average	0.2272	4.671	95.329

Figure 2. Microstructure of prepared Al/5%vol Grp-MMC at 100 X.

These figures also represent the actual percentage of reinforced particles present in the matrix. In case of Al/5%vol Al_2O_3-MMC actual percentage of Al_2O_3 is 4.975 with 95.025% Al-matrix. Similarly, for Al/5%vol Gr_p-MMC actual percentage of Gr_p is 4.671 with 95.329% Al-matrix.

Table 3 represents the L_{27} (3^{13}) orthogonal array, experimental results of micro hardness and tensile strength and their S/N ratios (dB) of Al/5vol% Al_2O_3-MMC and Al/ 5 vol% Grp MMC.

Table 3. Experimental results and S/N Ratio (dB) for Micro Hardness and Tensile Strength of Al/ 5 vol% Al_2O_3 MMC and Al/ 5 vol% Grp MMC

Exp No.	A	B		Al/ 5 vol% Grp MMC							Al/ 5% vol Al_2O_3-MMC			
				Micro Hardness(VHN)					Tensile Strength (MPa)		Micro Hardness (VHN)		Tensile Strength (MPa)	
				Y1	Y2	Y3	Avg	S/N ratio	Avg.	S/N ratio	Avg.	S/N ratio	Avg.	S/N ratio
1	750	150		67.6	67.0	67.3	67.3	36.56	191.37	45.64	135..1	43.87	274.25	49.23
2	750	150		67.2	67.9	67.7	67.6	36.59	192.6	45.64	137.3	43.88	275.56	49.26
3	750	150		68.2	68.9	68.1	68.4	36.70	193.17	45.72	136..4	43.90	276.24	49.31
4	825	225		70.7	70.8	70.5	70.7	36.98	195.16	45.81	138.6	43.05	278.32	49.37
5	825	225		71.2	71.0	71.4	71.2	37.04	195.49	45.82	139.6	43.09	278.61	49.40
6	825	225		71.7	71.5	71.3	71.5	37.08	196.58	45.87	137.3	43.11	279.54	49.46
7	900	300		72.2	72.5	72.4	72.4	37.19	196.82	45.88	135.5	43.14	279.81	49.47
8	900	300		72.1	72.6	72.7	72.5	37.20	198.51	45.95	136.4	43.11	280.32	49.49
9	900	300		72.9	72.0	72.8	72.6	37.21	198.87	45.97	135.4	43.21	281.75	49.50
10	825	300		65.4	65.6	65.3	65.4	36.31	189.64	45.56	137.3	43.71	272.55	49.17
11	825	300		66.2	66.8	66.4	66.5	36.45	190.42	45.59	138..5	43.79	273.60	49.20
12	825	300		65.2	65.0	65.1	65.1	36.27	189.33	45.54	139.2	43.66	272.26	49.15
13	900	150		69.2	69.5	69.6	69.4	36.83	193.74	45.74	134.5	43.02	277.25	49.33
14	900	150		70.2	70.0	70.3	70.2	36.92	194.54	45.78	133.7	43.04	277.62	49.35
15	900	150		68.8	68.6	68.7	68.7	36.73	195.84	45.84	132.5	43.95	276.54	49.32
16	750	225		69.7	69.8	69.3	69.6	36.85	194.89	45.79	139.9	43.07	277.47	49.34
17	750	225		71.6	71.4	71.3	71.4	37.07	188.86	45.53	141.5	43.11	278.84	49.42

18	750	225		70.4	70.6	70.3	70.4	36.95	188.64	45.51	136.7	43.04	278.16	49.35
19	900	225		64.8	64.9	64.6	64.8	36.23	188.42	45.50	139.6	43.66	271.82	49.14
20	900	225		64.1	64.5	64.4	64.3	36.17	190.63	45.60	141.9	43.60	271.52	49.12
21	900	225		64.2	64.0	64.3	64.2	36.14	189.84	45.57	136.3	43.60	271.26	49.11
22	750	300		66.9	66.7	66.3	66.6	36.47	190.63	45.60	139.6	43.83	273.91	49.21
23	750	300		65.9	65.7	65.5	65.7	36.35	189.84	45.57	142.2	43.73	272.81	49.17
24	750	300		66.0	66.5	66.1	66.2	36.42	190.17	45.58	136.5	43.76	273.26	49.19
25	825	150		69.4	69.3	69.1	69.3	36.81	193.56	45.74	136.6	43.98	276.85	49.32
26	825	150		67.1	67.5	67.4	67.3	36.56	191.81	45.66	141.5	43.94	274.73	49.24
27	825	150		68.4	68.0	68.5	68.3	36.69	192.85	45.70	136.1	43.93	275.85	49.27

Table 4. ANOVA and 'F' test for tensile strength and micro hardness of Al/5 vol% Al_2O_3-MMC

Parameters	D.o.f.	Tensile Strength of Al/5% vol Al_2O_3-MMC				Micro Hardness of Al/5 vol% Al_2O_3-MMC			
		Sum of square	Variance	'F' test value	% of contribu-tion	Sum of square	Variance	'F' test value	% of contribu-tion
X1	2	9.862	4.931	10.8	1.43	87.929	38.9645	29.91	16.04
X2	2	5.496	2.748	6.02	0.80	119.35	64.6749	49.65	21.77
X3	2	1.143	0.572	1.25	0.17	106.956	55.9779	42.98	19.51
X1.X2	4	642.224	160.556	351.73	93.12	163.206	19.8016	15.2	29.77
X1.X3	4	1.775	0.444	0.97	0.26	25.389	3.8473	2.95	4.63
X2.X3	4	0.856	0.214	0.47	0.13	28.695	4.6738	3.59	5.23
Error	62	28.301	0.456		4.09	16.519	1.3025		3.01
Total	80	689.657				548.115			100

X1: Pouring temperature; X2: Stirring speed; X3: Stirring time.

Table 5. ANOVA and 'F' test for tensile strength and micro hardness of Al/5 vol % Grp-MMC

Parameters	D.o.f.	Tensile Strength of Al/5% vol Grp-MMC				Micro Hardness of Al/5 vol% Grp-MMC			
		Sum of square	Variance	'F' test value	% of contribu-tion	Sum of square	Variance	'F' test value	% of contribu-tion
X1	2	122.086	61.043	61.23	16.45	5.45	2.725	8.28	0.99
X2	2	20.13	10.065	10.1	2.72	4.729	2.364	7.19	0.87
X3	2	1.117	0.559	0.56	0.15	0.349	0.174	0.53	0.08
X1.X2	4	486.466	121.616	121.99	65.54	511.159	127.79	388.54	93.5
X1.X3	4	35.997	8.999	9.03	4.85	2.255	0.564	1.71	0.42
X2.X3	4	14.62	3.655	3.67	1.97	2.242	0.561	1.7	0.41
Error	62	61.812	0.997		8.32	20.392	0.329		3.73
Total	80	742.227			100	546.576			100.00

X1: Pouring temperature; X2: Stirring speed; X3: Stirring time.

Table 4 represents ANOVA and 'F' test values for tensile strength and micro hardness of the prepared Al/ 5 vol% Al_2O_3-MMC. From the Table 4, it is clear that the stirring speed and stirring time are the most significant and significant parameters with 'F' test value 49.65 and 42.98 respectively for micro hardness of Al/ 5 vol% Al_2O_3-MMC. Pouring temperature is the third influencing factor i.e. comparatively third significant factor with 'F' test value 29.91 on the micro hardness of prepared Al/5 vol% Al_2O_3-MMC composite by stir cast method. From the Table 4, it is clear that interaction of pouring temperature and stirring speed has great influence with 93.12% contribution on tensile strength of Al/5% vol Al_2O_3-MMC. Pouring temperature is the second influencing factor with 10.80 'F' test value on the tensile strength of Al/5% vol Al_2O_3-MMC composite prepared by stir cast method.

Table 5 represents ANOVA and "F" values for tensile strength and micro hardness of Al/5 vol % Gr_p-MMC cast sample. From the Table 5, it is clear that the interaction of pouring temperature and stirring speed has great influence with 65.54% and 93.5% contribution on tensile strength and micro hardness of Al/ 5 vol% Gr_p-MMC respectively. Pouring temperature is the second influencing and significant factor with 8.28 'F' test value on the micro hardness of the prepared Al/5% vol Gr_p-MMC sample. Pouring temperature is the second influencing and significant factor with 16.45 % contribution on the tensile strength of prepared Al/5 vol % Gr_p-MMC by stir cast method.

The average S/N ratio (dB) by factor level for micro hardness and tensile strength of prepared Al/ 5 vol% Al_2O_3 and Al/ 5 vol% Gr_p-MMC,s are displayed graphically in figures 3 and figure 4 respectively.

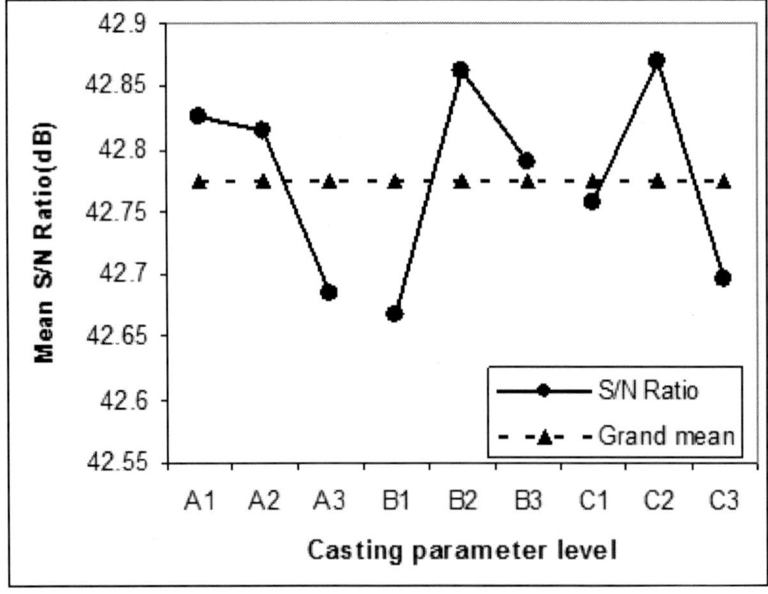

Figure 3. (a) S/N ratio(dB) graph for Micro hardness of Al/ 5 vol% Al2O3-MMC.

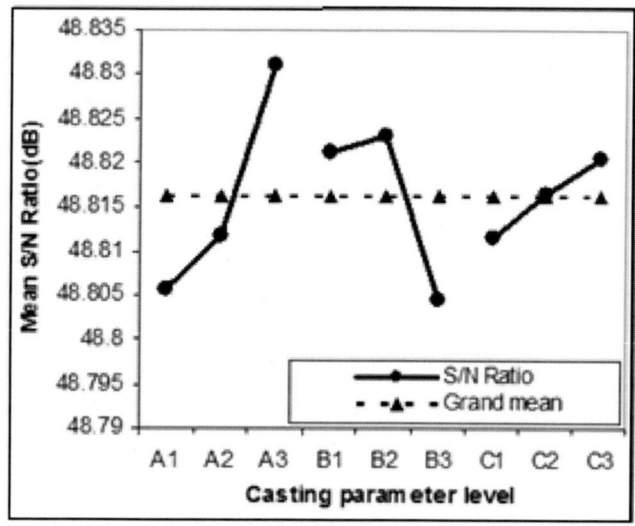

Figure 3. (b) S/N ratio(dB) graph for Tensile of Al/ 5 vol% Al2O3-MMC.

From the figure 3 (a), it is concluded that for maximum micro hardness the optimal parametric setting is $A_1B_2C_2$ i.e. at 750^0C pouring temperature, 225 r.p.m stirring speed and 6 min stirring time. From figure 3(b), it is concluded that for maximum tensile strength the optimal parametric setting is $A_3B_2C_3$ i.e. at 900^0C pouring temperature, 225 r.p.m. stirring speed and 9 min stirring time.

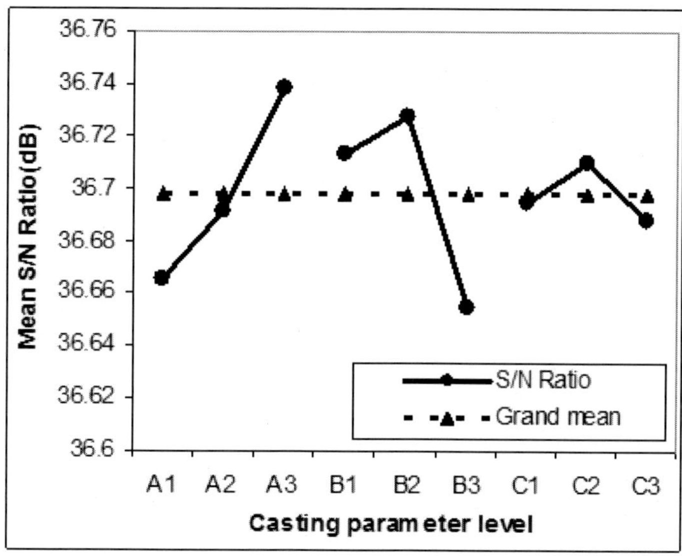

Figure 4. (a) S/N ratio(dB) graph for Micro hardness of Al/ 5 vol% Grp-MMC.

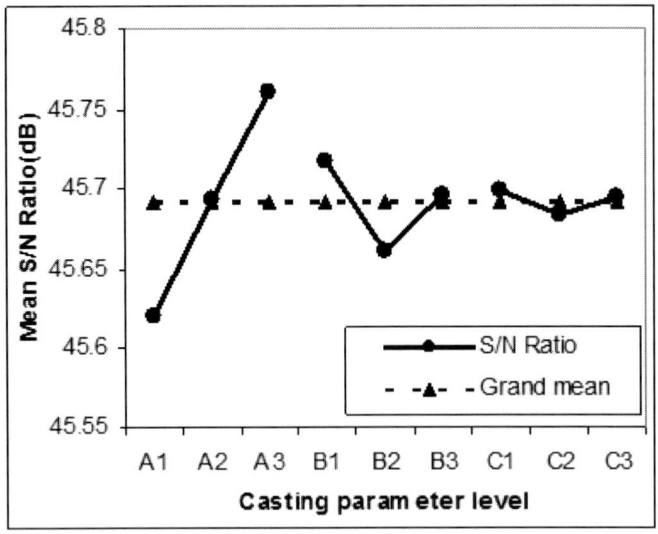

Figure 4. (b) S/N ratio(dB) graph for Tensile of Al/ 5 vol% Grp-MMC.

Similarly, for Al/ 5 vol% Gr_p-MMC; from the figure 4(a), it is concluded that for maximum micro hardness the optimal parametric setting is $A_3B_2C_2$ i.e. at 900^0C pouring temperature, 225 r.p.m. stirring speed and 6 min stirring time. From the figure 4(b), it is concluded that for maximum tensile strength, the optimal parametric setting is $A_3B_1C_1$ i.e. at 900^0C pouring temperature,150 r.p.m. stirring speed and 3 min stirring time.

4. Mathematical Models for Micro Hardness and Tensile Strength of Cast MMC

The Gauss elimination method is utilized and developed the mathematical models for micro hardness and tensile strength of the stir cast Al/ 5 vol% Al_2O_3-MMC and Al/ 5 vol% Gr_p-MMC samples. The mathematical model for Micro Hardness of prepared Al/ 5 vol% Al_2O_3-MMC is

$$Y_{\text{micro Al/ 5 Vol\%-Al2O3}} = -301.17284 + 28.2592\ X1 + 21.130864\ X2 + 294.10493X3 - 0.003735\ X1.X2 - 0.0061728\ X1.X3 + 0.012345\ X2.X3 - 0.00174485\ X1^2 - 0.037399X2^2 - 25.75189\ X3^2 \qquad \text{Eqn.1}$$

$R^2 = 0.85$

Where, X1, X2 and X3 are the pouring temperature, stirring speed and stirring time respectively.

The mathematical model for tensile strength of prepared Al/5% vol Al_2O_3-MMC is

$Y_{\text{tensile Al/ 5\% vol-Al2O3}}$ = 337.72719 - 0.10819 X1 - 0.18071 X2 + 0.41972222 X3 + 0.0002450 X1.X2 - 0.0005049 X1.X3 + 0.00026173 X2.X3 + 0.00003730 $X1^2$ - 0.00058798$X2^2$- 0.001131 $X3^2$ Eqn. 2

$R^2 = 0.958$

The variation of micro hardness with pouring temperature and stirring time for the prepared Al/5%vol Al_2O_3-MMC cast sample is shown in Figure 5(a). In this graph along the X-axis '0' represents 725^{0}C of pouring temperature and '10' represents 925^{0}C pouring temperature. Similarly, along the Y-axis '0' represents the 0 min of stirring time and '10' represents the 9.5 min of stirring time. From the Figure 5(a), it is clear that at moderate pouring temperature and moderate stirring time micro hardness is comparatively more i.e. at interaction of 825^{0}C pouring temperature and 6 min of stirring time. Similarly, the variation of tensile strength with pouring temperature and stirring time for the prepared Al/5%vol Al_2O_3-MMC cast sample is shown in Figure 5(b). From the Figure 5(b), it is clear that at high pouring temperature and high stirring time (i.e. at high interaction of 900^{0}C pouring temperature and 9 min of stirring time) tensile strength is comparatively high.

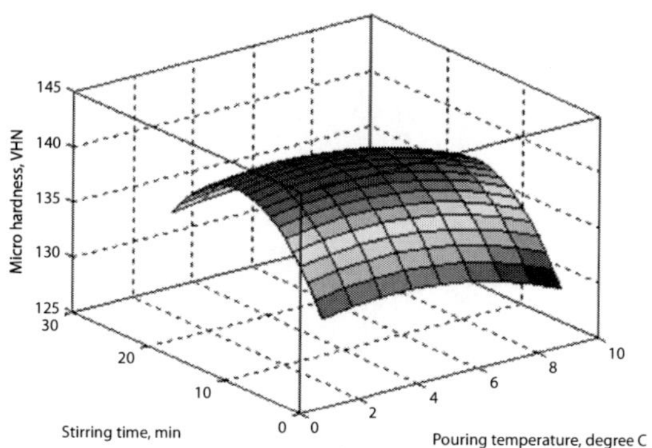

Figure 5. a)Variation of micro hardness with pouring temperature and stirring time for prepared Al/ 5 Vol% Al_2O_3 – MMC.

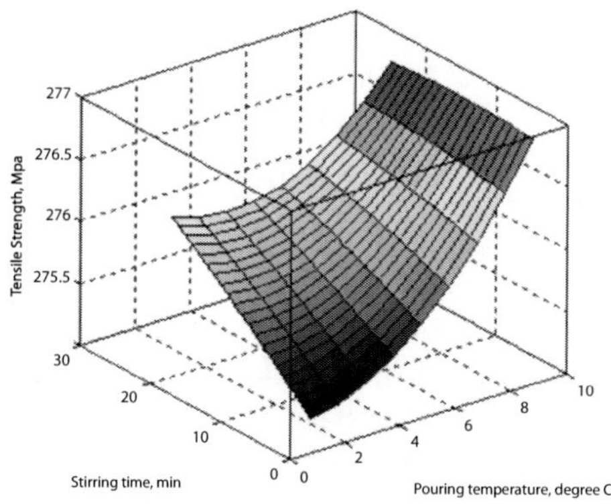

Figure 5. (b)Variation of tensile strength with pouring temperature and stirring time for prepared Al/ 5 Vol% Al2O3 – MMC.

The mathematical model for Micro Hardness of prepared Al/ 5 vol% Gr_p-MMC as follows

$Y_{micro\ 5\ Vol\%\text{-}Gr\ p}$= 106.578703 - 0.06379 X1 - 0.141592 X2 + 0.98518X3 + 0.00020493 X1.X2 - 0.00097531 X1.X3 + 0.00001235 X2.X3 + 0.000016790 $X1^2$ – 0.0000672 $X2^2$ - 0.015432 $X3^2$ Enq. 3

R^2 = 0.96

The mathematical model for tensile strength of prepared Al/ 5 vol% Gr_p-MMC as follows

$Y_{Tensile\ 5\ Vol\%\text{-}Grp}$= 234.1566 -0.02166 X1 -0.2767 X2 -3.32765X3+ 0.0002458 X1.X2 + 0.003885 X1.X3 - 0.00071235X2.X3 - 0.00002238 $X1^2$ – 0.00016506$X2^2$ - 0.025205 $X3^2$ Eqn. 4

R^2 = 0.916

CONCLUSION

The processing of Al/Gr_p and Al/Al_2O_3 can be possible by stir casting process. For maximum micro hardness and tensile strength of Al/5vol% Al_2O_3-MMC the optimal parametric combinations are $A_1B_2C_2$ and $A_3B_2C_3$ respectively. For maximum micro hardness and tensile strength of Al/5vol% Grp-MMC, the optimal parametric setting is $A_3B_2C_2$ and $A_3B_1C_1$ respectively. The stirring speed and stirring time are the most significant and significant parameters with 'F' test value 49.65 and 42.98 respectively for micro hardness of the prepared Al/ 5 vol% Al_2O_3-MMC. Where as the interaction between pouring temperature and stirring speed has great influence with 93.5% and 65.54% contribution on the micro hardness and tensile strength of the prepared Al/5 vol% Grp-MMC sample respectively. The developed mathematical models for micro hardness and tensile strength of the cast samples are successfully proposed for proper evolution of casting parameters during stir casting of MMC,s.

REFERENCES

Andreasen, J. L. and De Chiffre, L.;(1993),*Automatic Chip Breaking Detection in Turning by Frequency Analysis of Cutting Force,* Annals of CIRP, Vol. 41/1; pp.45-48.

Hashim,J.; Looney, L.; and Hashmi, M.S.J.;(1999), *Metal matrix composites: production by stir casting method,* J of Materials Processing Technology; 92-93, pp. 1-7.

Hashim,J.; Looney, L.; and Hashmi, M.S.J.;(2002), *Particle distribution in cast metal matrix composites Part I* J. of Materials Processing Technology; 123, pp. 251-257.

Hunt(Jr.), Warren H; Osman, Todd M. and Lewandowski, John J.;(1993),*Micro and Macrostructural Factors in DRA Fracture Resistance*, JOM, January 1993, PP. 30-35.

Manoharan,M.; and Gupta,M.;(1999) *Effect of silicon carbide volume fraction on the work hardening behaviour of thermomechanically processes aluminium based metal matrix composite;* J. of Composites Part B 30, pp. 107-112.

Montgomery,D.C; (1997), *Design and Analysis of Experiments,* Wiley, New York

Naher, S.; Brabazon,D.; Looney;L.;(2003), *Simulation of the stir casting process,* J. of Materials Processing Technology; 143-144; pp. 567-57.

Nai,S.M.L.; and M Gupta,M.; (2003*),Syntesis and characterization of free standing, bulk Al/SiCp functionally gradient materials: effects of different stirrer geometries* Materials Research Bulletin, 38, pp.1573-1589.

Ourdjini,A.; Chew,K.C.; Khoo;B.T.;(2001), *Settling of silicon carbide particles in cast metal matrix composites;* J. of Materials Processing Technology; 116, pp.72-76.

Seo,Y.H.; and Kang,C.G.;(1999*),Effects of hot extrusion through a curved die on the mechanical properties of SiCp/Al composites fabricated by melt stirring*, J. of Composites Science and Technology; 59, pp. 643-654.

Zhou,Z, W.; and Xu, M.;(1997*),Casting of SiC Reinforced Metal Matrix Composite;* J. of Materials Processing Technology; 63, pp.358-363.

In: Metal Matrix Composites
Editor: J. Paulo Davim

ISBN: 978-1-61209-771-8

Chapter 5

AN OVERVIEW ON THE RESEARCH OF IN-SITU TITANIUM MATRIX COMPOSITES

Weijie Lu, Liqiang Wang**, *Jining Qin and Di Zhang
State key laboratory of metal matrix composites,
Shanghai Jiaotong University, Shanghai, China

ABSTRACT

In-situ titanium matrix composites have drawn great attention due to the superior properties over titanium alloys. Recent research achievements of synthesizing methods, selection of matrix and reinforcements, microstructure, mechanical properties and superplastic behavior are reviewed. The present problems are put forward and the further research fields are also discussed.

1. INTRODUCTION

Titanium matrix composites (TMCs) are composite materials defined as titanium or titanium alloy implanted hardened ceramic reinforcement. It combines the excellent ductility and toughness of titanium with high strength and high modulus of ceramic, which gives the materials higher shear strength, compressive strength and better high-temperature mechanical properties. TMCs, which have attractive physical and mechanical properties, such as high modulus, high strength, oxidation resistance, has been proven in many studies [1-2].

Titanium matrix composites are divided into two categories: continuous fiber reinforced titanium matrix composites and particle reinforced titanium matrix composites. The main area in earlier researches was titanium matrix composites reinforced with silicon carbide fiber, which can improve the mechanical properties of the matrix alloy significantly. However, fiber-reinforced titanium matrix composites are subject to the following factors: high price, complex process, anisotropy, interface reaction. Therefore, the study of discontinuously

*Corresponding author, E-mail: wang_liqiang@sjtu.edu.cn

reinforced titanium matrix composites is one of the most important research fields [3-5]. In addition, the ceramic reinforcement can improve significantly the wear resistance of the matrix alloys [6-7], which can be used in aerospace and military fields to meet the requirements of wear resistance and corrosion resistance.

The preparation techniques of discontinuously reinforced titanium matrix composites can be divided into external and in situ synthesis methods. In traditional external techniques, the ceramic reinforcement which is pre-prepared is added into the titanium matrix in the state of power. Thus, the scale of the reinforcement in the traditional synthesis is limited by the initial particle size to micron or millimeter. In addition, there are some disadvantages for traditional external techniques such as interface saturation, interface reaction between reinforcement and matrix, and high price.

In order to solve the problems in traditional external techniques, in situ synthesis technology is utilized to prepare discontinuously reinforced titanium matrix composites. Reinforcement is produced by the chemical reaction. Using in situ synthesis technology, the compatibility between the reinforcement particles and matrix is improved and the interface reaction can also be avoided. What is more, the system between reinforcing particles and matrix is stable in thermodynamics. The characteristics of in situ synthesis technology mentioned above are very useful to promote the service temperature of titanium matrix composites. Nowadays, in situ synthesized titanium matrix composite is an important research area in composites [3].

2. Matrix and Reinforcement

2.1 Matrix

In accordance with the mixing law, the choosing of matrix is very important to the mechanical properties of composites. Ti-6Al-4V which has good comprehensive performance is widely used as the matrix material in discontinuously reinforced titanium matrix composites. In the aviation and aerospace fields, near α and $\alpha + \beta$ type alloys are chosen as the matrix materials catering to the requirements of high-temperature strength and creep resistance.

2.2 Reinforcement

For in situ titanium matrix composites, high melting point and high hardness phases are usually used as reinforcement materials, such as metal, non-metallic compound and intermetallics. One of the important physical parameters of reinforcement particles is the thermal expansion coefficient. As for titanium alloy with high yield strength, high thermal residual stress could be avoided. Therefore, matrix and reinforcement particles should be similar in the thermal expansion coefficients. In addition, the chemical compatibility between matrix and reinforcement is also very important for the interface reaction happening in high temperature, which reduces the interface strength strongly.

Physical properties of reinforcements used in titanium matrix composites are shown in Table 1. Under given conditions, the addition of SiC、Al_2O_3 and Si_3N_4 will result in serious interfacial reactions with active Ti element [8-13], therefore, SiC、Al_2O_3 and Si_3N_4 are not perfect reinforcements. Because of the instability of B_4C、TiB_2 and ZrB_2 in titanium matrix, TiC and TiB will form during preparing process. However, TiC and TiB possess very high melting temperatures, and are stable in titanium matrix. In addition, TiC and TiB have good compatibility with titanium matrix, which can exist without any interfacial reactions [14-15]. Moreover, TiC, TiB and titanium matrix also own similar poisson's ratios and close density. Thus, if the difference of thermal expansion coefficients among them is controlled below 50% (Thermal expansion coefficient of titanium is $9\sim10.8\times10^{-6}$/K), the thermal residual stress during preparation can be decreased evidently. Furthermore, since elastic modulus of TiB or TiC is 4 to 5 times bigger than titanium, the mechanical properties of matrix can be increased largely. So TiB and TiC are more ideal reinforcements for discontinuously reinforced titanium matrix composite.

Table 1. Properties of reinforcements used in titanium matrix composites

Particle	Density ($g.cm^{-3}$)	Melting point (K)	Coefficient of thermal expansion ($10^{-6}K^{-1}$)	Modulus (GPa)
SiC	3.19	2970	4.63 (25~500°C)	430
TiC	4.99	3433	6.52~7.15 (25~500°C)	440
B_4C	2.51	2720	4.78 (25~500°C)	445
TiB_2	4.52	3253	4.6~8.1	500
ZrB_2	6.09	3373	5.69 (25~500°C)	503
TiB	4.05	2473	8.6	550
Al_2O_3	4.00	2323	8.3	420
Si_3N_4	3.20	2173	2.5	385

In addition to that mentioned above, rare earth oxide is considered as the most hopeful reinforcements for titanium alloy. The rare earth elements which could be added in titanium alloy are La、Nd、Y、Ce、Er、Gd and so on. The rare earth oxide is stable and owns very high melting temperature, which plays mainly role in internal oxidation after added into titanium matrix. The rare earth oxides distribute diffusely over titanium matrix, which can strengthen the matrix alloy. Hence, the addition of rare earth element promotes the instantaneous strength and rupture strength of matrix alloy at high temperature obviously [16-17]. Moreover, the rare earth element is also beneficial to refine the grain of matrix and increase the thermal stability [18].

3. Preparation

Nowadays, particle reinforced titanium matrix composites could be in situ synthesized by the following methods: powder metallurgy (P/M) [19-24], mechanical alloying (MA) [25-26], self-propagating high temperature synthesis (SHS) [27-29], exothermic dispersion (XDTM) [30], rapid solidification processing(RSP) [31-32] and a variety of melting and casting techniques [33-36]. Among them, the powder metallurgy is considered as the most popular method to prepare in situ titanium matrix composites. However, the main disadvantages of this method are complex processing, difficulty in manufacturing large parts and productions. The reactant and the matrix alloy are melted using the traditional melting process. Figure 1 shows the traditional preparing technologies for discontinuously reinforced titanium matrix composites. The reinforcements can be in situ synthesized during the melting process. By the ordinary casting method without changing the original titanium alloy melting process and equipments, the costs can also be reduced greatly. Figure 2 shows the process of preparing in situ synthesis of titanium matrix composites. In this method, during the preparation of TMCs, reinforcements are appearing and growing in the matrix without special equipment and technology. The key issues are to control the type, shape, size, content and distribution of reinforcements in in situ synthesis of titanium matrix composites.

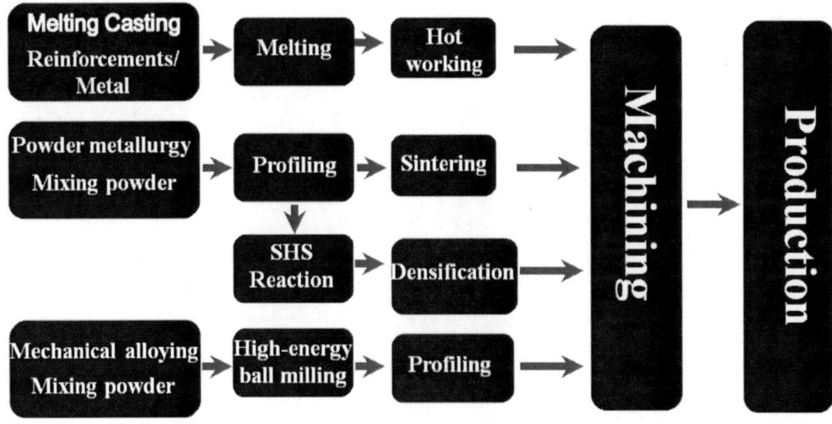

Figure 1. Traditional preparing technologies for discontinuously reinforced titanium matrix composites.

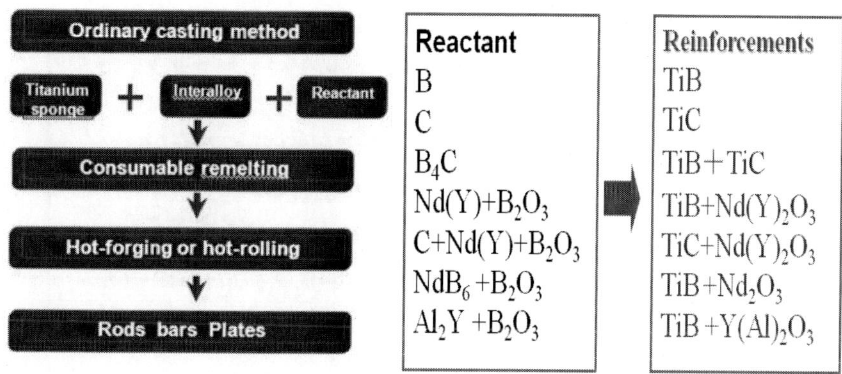

Figure 2. Processing of preparing in-situ synthesis of titanium matrix composites.

4. In Situ Reaction and Thermodynamics

Nowadays, in situ synthesized titanium matrix composites are produced using the following reaction formula:

$$Ti + B \rightarrow TiB \quad (1)$$

$$Ti + C \rightarrow TiC \quad (2)$$

$$5Ti + B_4C \rightarrow 4TiB + TiC \quad (3)$$

$$14Ti + 2REB_6 + B_2O3 = 14TiB + RE_2O_3 \quad (4)$$

Where RE denotes rare earth element. Zhang and Lu et al [37-39] have used traditional equipment to melt TiC / Ti, TiB / Ti and TiB + TiC / Ti composites and have calculated reaction Gibbs free energy and reaction enthalpy between titanium and graphite, boron, boron carbide. It was found that Gibbs free energy of each reaction and the reaction enthalpy are negative at high temperatures, as shown in figure 3. So it can be known that the reactions (1-3) were exothermic and thermodynamically feasible.

The simple substance in raw materials is very reactively to react with each other in high temperature, which affects the metallurgical results. Therefore, in situ titanium matrix composites are developed with a trend of diversification and multiple scales. In situ synthesized TiB, TiC and rare earth oxides (RE_2O_3) multiple ceramic particulates reinforced titanium matrix composites have been produced by Yang[40] ,using the reaction between titanium and B_4C, REB_6 or B_2O_3. According to the process control, nano rare earth oxides particles can disperse in the matrix equably.

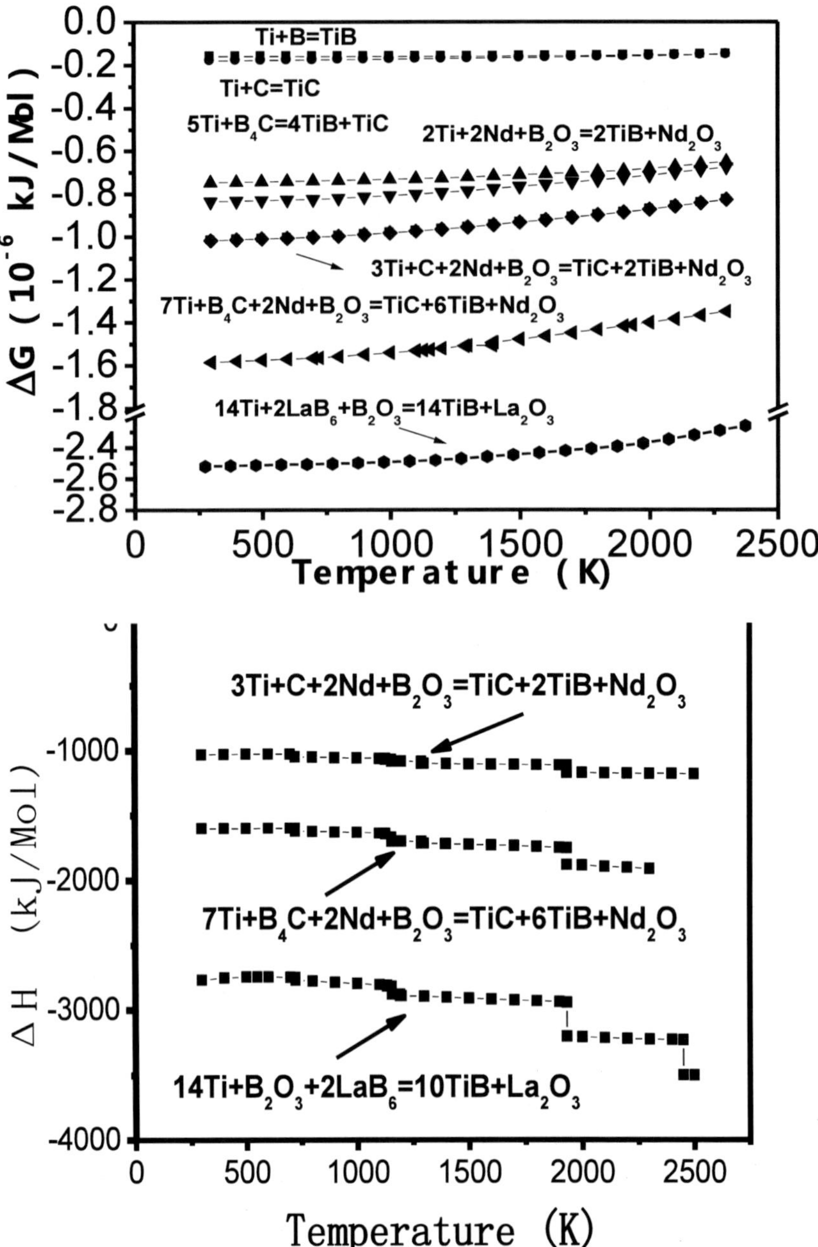

Figure 3. Gibbs energy and enthalpy of in situ reactions.

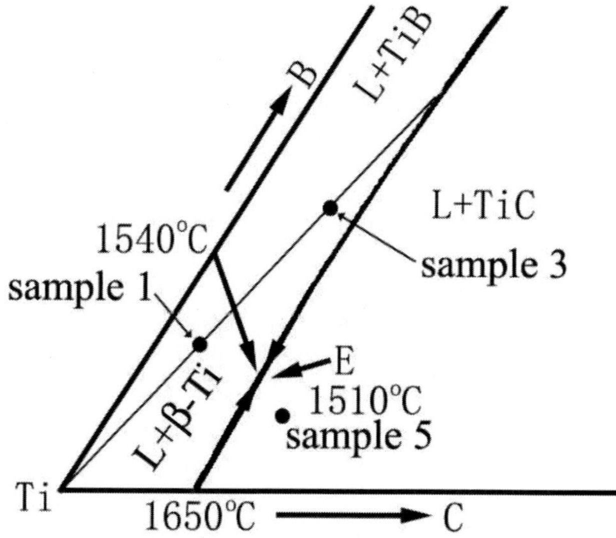

Figure 4. Liquidus surfaces of Ti-B-C in Ti-rich range.

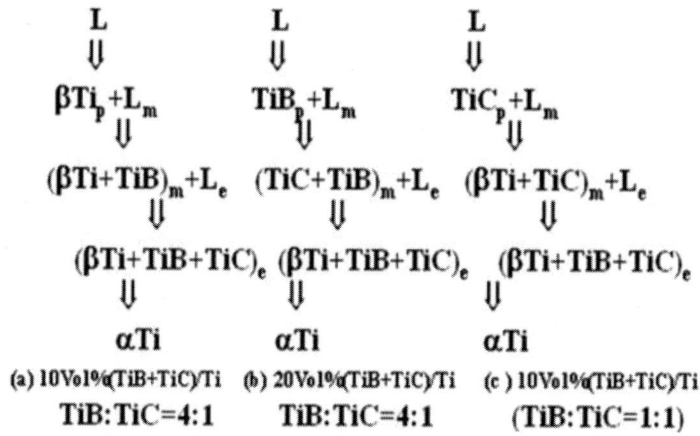

Figure 5. Solidify process of in situ synthesized titanium matrix composites.

According to the liquidus surfaces of Ti-B-C in Ti-rich range shown in figure 4. In order to control the reinforcement morphology, we designed the content of additions to obtain the reinforcements with proper size and morphology. Figure 5 shows the solidify process of in situ synthesized titanium matrix composites with different additions.

5. MICROSTRUCTURE

In titanium matrix composites, the microstructure of TiC reinforcements diversifies micro-structure characteristics, such as spherical, equiaxed grain, short stick, and feathery. The different kinds of shape and size of reinforcements have close relationship with preparation methods, solidification rate, carbon content and alloying elements. During combustion assisted cast (CAC) process, TiC takes usually the form of three-dimensional dendrite. Lu [41] pointed out that in the casting process of titanium matrix composites, the forming of dendritic TiC is mainly attributed to the ingredients in the solidification process, as shown in figure 6. It can be observed that TiC reinforcement tends to be smaller and grows into equiaxed particles.

TiB reinforcement is observed as whisker, needle, sheet and tube. In addition, the interface between TiB and titanium matrix is good and there is a certain crystallographic orientation relationship between them. As shown in figure 7, with a B27 structure, the cross section of TiB whiskers is hexagonal. Because the rate of crystal growth along [010] direction is much faster than others, the reinforcements are more slender with a higher aspect ratio, which is very important for composites. Moreover, Lu [42] has also found that there is a certain crystallographic orientation relationship between TiB whiskers and titanium matrix. Primary sheet TiB phase appears in the composites with larger amount of reinforcement [43]. The growth direction of the hollow tube is the same as that of needle-like TiB, which is along the [010] direction. The appearance of tubular TiB is attributed to the higher growth rate of outer surface resulted from undercooling along [010] direction in substance.

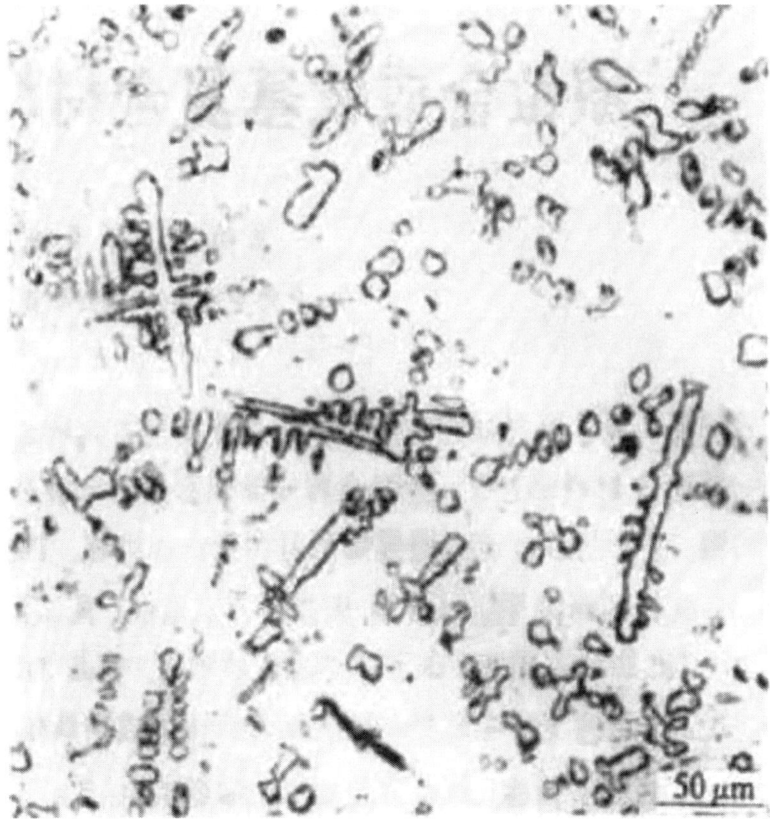

Figure 6. Morphology of TiC.

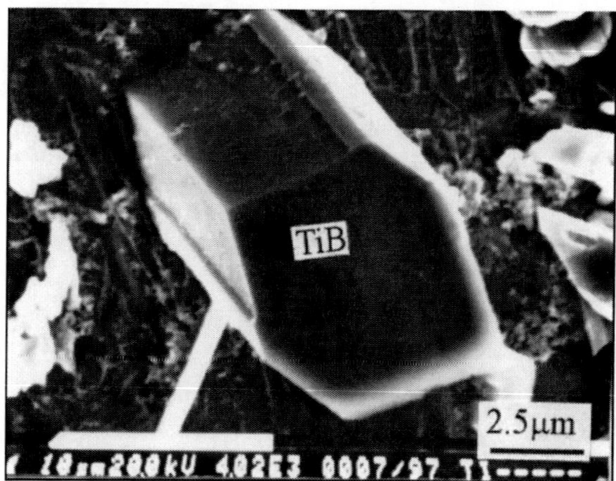

Figure 7. Typical SEM morphology of TiB whisker.

When the content of the rare earth elements is low, because of the geometrical symmetry of the crystal structure, rare earth oxides in the titanium alloy are generally observed as spherical shape [44-46]. In addition, as for many rare earth elements, the solubility in α titanium alloys is lower than that in β titanium alloys. During solidification process, there are many nano-scale secondary precipitation of rare earth oxide particles, distributing over the titanium matrix [47], as shown in figure8. When the rare earth content is high, due to the impact of undercooling, the oxide is dendritic. Certain relationships among Y_2O_3, Nd_2O_3 and TiB reinforcement in crystallographic orientation have also been found [47-49].

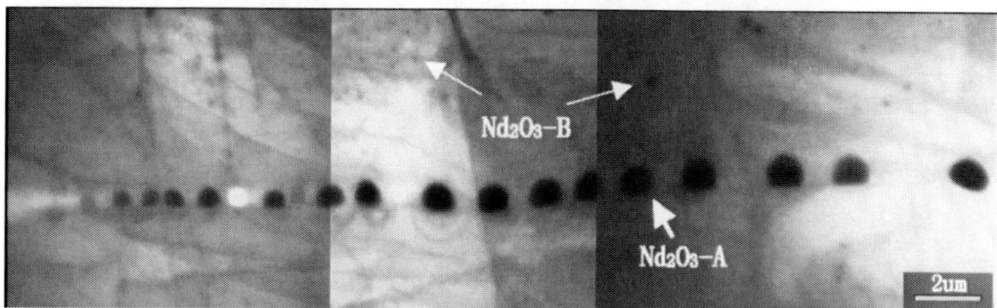

Figure 8. TEM morphology of Nd_2O_3 particles.

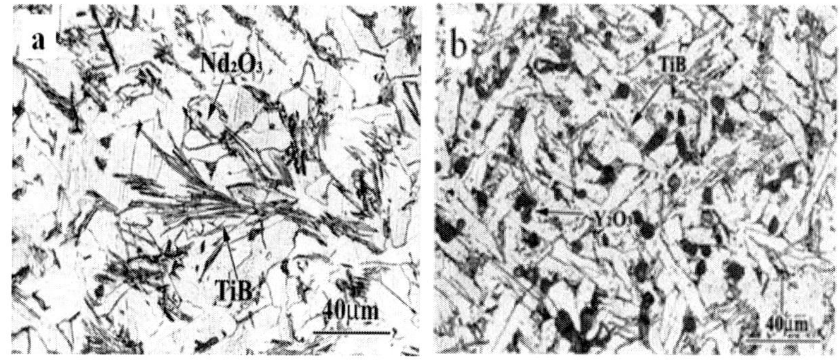

Figure 9. (Continued).

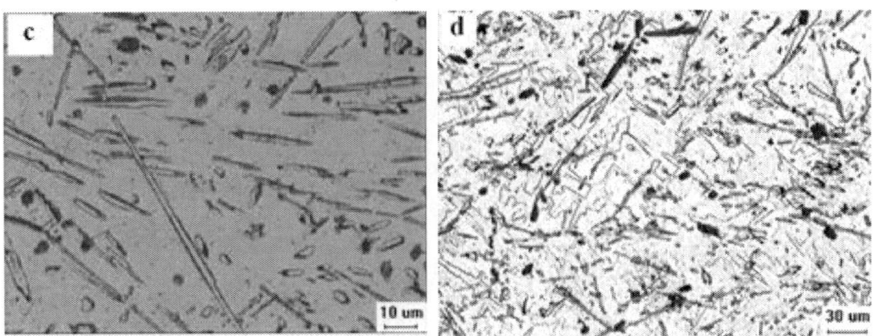

Figure 9. OM morphology of TMCs: (a) (TiB+Nd2O3)/Ti; (b) (TiB+Y2O3)/Ti; (c) (TiB+Ld2O3)/Ti; (d) (TiB+TiC+Y2O3)/Ti.

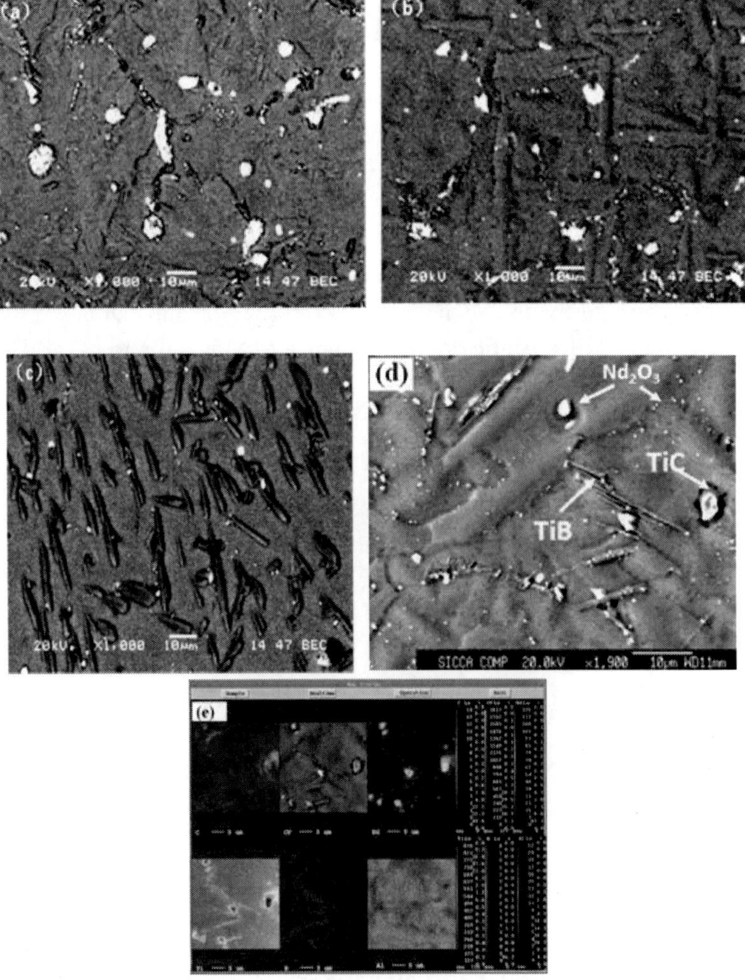

Figure 10. Backscatter scanning photos of TMCs corresponding second-electron images of TMCs (a)Nd (b)Al2Nd (c)LaB6 (d) TiB+TiC+Nd2O3 (e) corresponding second-electron images of (d).

Figure 9 shows the OM morphology of TMCs with different reinforcements. As shown in figure 9, reinforcements such as TiB, Nd2O3, Y2O3, Ld2O3 and TiC are observed clearly. Figure 10 shows the backscatter scanning photos of TMCs corresponding second-electron images of TMCs. The size and distribution reinforcement can also be observed. Reinforcements distribute equably as the backscatter scanning morphologies shown in figure 10.

Figure 11 shows the high-resolution transmission electron microscopy of TMCs. The main orientation relationships between titanium matrix and reinforcements are studied. As shown in figure 11, certain relationship between titanium matrix and reinforcements, reinforcements and reinforcements are discovered. Table 2 shows the main orientation relationships between titanium matrix and reinforcements.

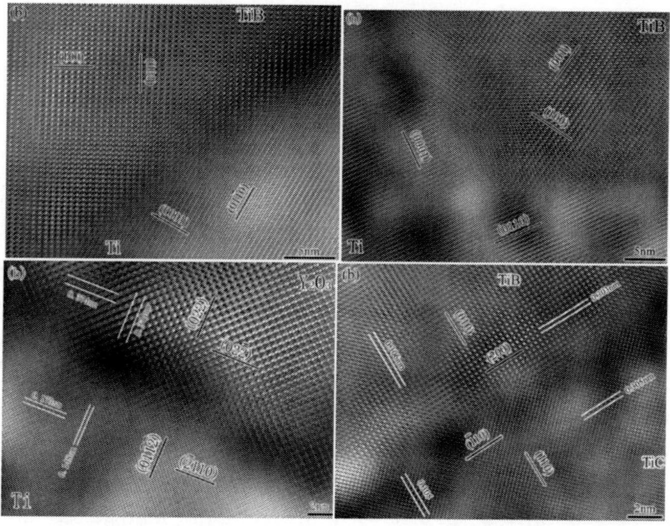

Figure 11. *High-resolution* transmission electron microscopy (HRTEM) of TMCs.

Table 2. Main orientation relationships between titanium matrix and reinforcements

Titanium/Reinforcements	The main orientation relationships
Ti and TiB	[001]TiB//[0001]Ti ; (020)TiB//(2-1-10)Ti, (200)TiB//(1-100)Ti [010]TiB//[11-20]Ti ; (001)TiB//(0001)Ti, (200)TiB//(1-100)Ti
Ti and Nd_2O_3	[110]Nd_2O_3//[-12-1-3]Ti;(1-11)Nd_2O_3//(1-10-1)Ti, (001)Nd_2O_3//(2-1-10)Ti
Ti and Y_2O_3	[100] Y_2O_3 //[01-11]Ti, (044) Y_2O_3 //(01-1-2)Ti, [110] Y_2O_3 //[1-21-3]Ti, (-223) Y_2O_3 //(01-1-1)Ti ; [010] Y_2O_3 //[11-20]Ti ; (400) Y_2O_3 //(1-100)Ti ;
TiB and TiC	[001]TiB//[001]TiC, (010)TiB//(110)TiC, (200)TiB//(-110)TiC ; [010]TiB//[110]TiC, (001)TiB//(001)TiC, (200)TiB//(-110)TiC ;
TiB and Nd_2O_3	[011]TiB//[021] Nd_2O_3, (01-1)TiB//(-100) Nd_2O_3 TiB and Y_2O_3:
TiB and Y_2O_3	[-101]TiB//[021] Y_2O_3,(111)TiB//(1-21) Y_2O_3

6. Mechanical Properties

Comparing with the titanium alloy, both of the strength and elastic modulus of titanium matrix composites with TiB whisker and TiC particle are higher and increase with the reinforcement content.

Lu [50] has studied the mechanical properties of the Ti6242 alloy enhanced by TiB and TiC. It was found that the addition of equiaxed TiC particles improved the mechanical properties of composites. Yang [51] has studied the mechanical properties of multiple enhancements (TiB + TiC + La_2O_3) / Ti composites. Comparing with the (TiB + TiC) / Ti composites, higher tensile strength and plasticity were obtained in (TiB + TiC + La_2O_3) / Ti composite. The strengthening mechanism at room temperature was mainly attributed to the bearing of reinforcement, grain refinement of matrix alloy and the dispersion strengthening of La_2O_3 nano-particles. Geng [52] studied the high-temperature tensile properties of pure titanium-based in situ (TiB + Y_2O_3) / Ti composites and found that the tensile strength and yield strength are 2 to 3 times higher than that of the pure titanium under the same conditions. The strength of composites increased with the increase of volume fraction of Y_2O_3. Yang [53] has studied the high temperature mechanical properties of multiple multi-scale reinforced titanium matrix composites. With the increase of test temperature, the strength decreased and elongation increased. In addition, the fracture mechanisms of titanium matrix composites at high temperature were obviously different with that at room temperature. At room temperature the bearing fracture of TiB short fibers appeared mostly, however, at high temperature the interface debonding between the reinforcements and matrix occurred mainly.

Xiao has reported the high temperature mechanical properties of heat-resistant titanium matrix composites with reinforcements of TiB, TiC and La_2O_3 [54-55]. As shown in figure 12, the high temperature strength is higher than that of the matrix alloy and other high temperature titanium alloy. Compared with IMI834 and Ti1100 the high temperature ultimate strength of TMCs are improved greatly. The fracture mechanism at high temperature is different with that at room temperature. Figure shows the high temperature fracture mechanism of TMCs. As shown in figure 13(a),for the specimen with low content of reinforcements, a large number of fractured TiB short fibers appeared after high temperature tensile at 700 °C,which is similar to the result of room temperature tensile. When the content of TiB is 5%, both fractured TiB short fibers and interfacial debonding was observed when the high temperature tensile was carried out at 650°C(figure13(b)). With the increase of temperature, interfacial debonding was the major fracture mechanism, as shown in figure 13 (c). For the higher content of reinforcement, interfacial debonding contributed most when the temperature is 650 °C, as shown in figure 13 (d). In addition, high-temperature tensile properties are very sensitive to strain rate. When the strain rate decreases, high temperature tensile strength decreases and elongation increases.

Lu [56] has studied the rupture properties of Ti6242 alloy with reinforcements of TiB and TiC. Comparing with the matrix alloy, reinforcement improves the creep fracture properties of materials. Dislocation pileup appears and crack appears at the end of the TiB short fibers. Finally, the growing of the crack occurs, which is very deleterious to the material. The addition of graphite accelerates the appearance of more equiaxed TiC reinforcement, which is helpful to improve the creep properties of composites.

Xiao [57-58] has reported the high temperature creep behavior of multiple enhanced heat-resistant titanium matrix composites. The creep rate of composite is lower than that of IMI834 alloy by 1 to 2 times, and the addition

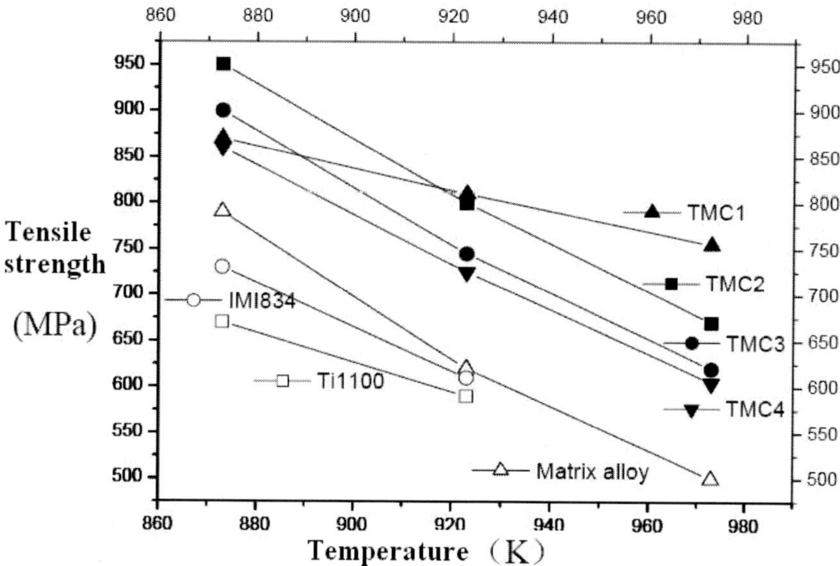

Figure 12. Temperature dependence of strength in TMCs. TMC1 : 2.8%vol.($TiB+La_2O_3$)/Ti, TMC2 : 5%vol.($TiB+TiC+La_2O_3$)/Ti. TMC3 : 10%vol.($TiB+TiC+La_2O_3$)/Ti, TMC4 : 10%vol.(TiB+TiC)/Ti.

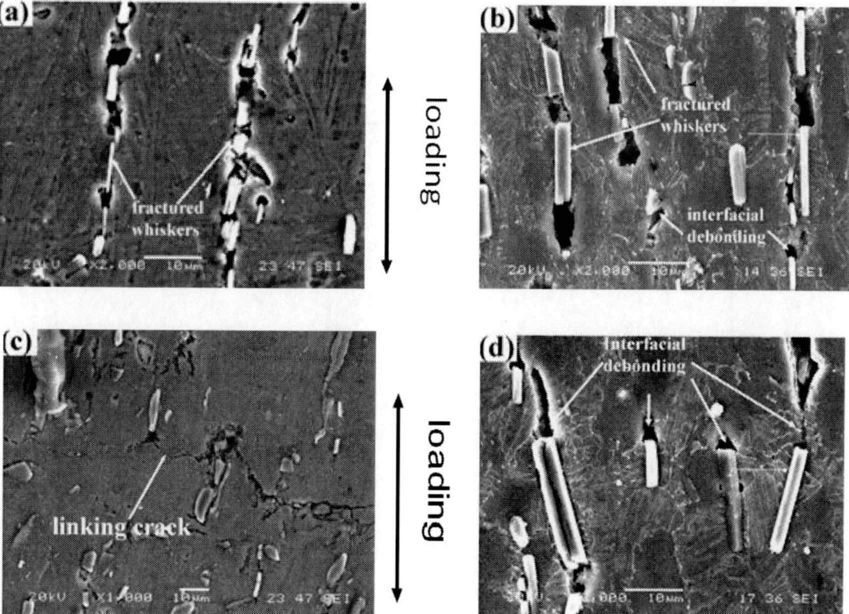

Figure 13. High temperature fracture mechanism of TMCs: (a) TiB 2.8%, 700°C,(b) TiB 5%, 650°C, (c)TiB 5%, 700°C, (d) TiB 10%, 650°C.

of reinforcement improves the creep resistance of composites significantly. Low stress and high stress areas appear in both matrix alloy and composite material at different stress. At lower stress and high stress area, the stress index of matrix alloy is 2 and 4.5, respectively. The creep resistance of composites is mainly based on the threshold stress and stress transfer effects. The improvement of the threshold stress caused by La_2O_3 particles is mainly observed in the high stress area and the threshold stress depends mainly on the dispersion of the reinforcement. Furthermore, the effect of stress transfer depends mainly on the aspect ratio of TiB short fibers. With the decreasing of the volume fraction of the reinforcement, the transfer factor at the low stress region is much higher than that at high stress region. The microstructure of reinforcement also plays an important role in the creep resistance of composites.

7. Oxidation Behavior

The oxidation behaviors of the in situ TiB / Ti, TiC / Ti and (TiB + TiC) reinforced titanium composites have been studied [59-62]. The reinforcements of TiB and TiC accelerate the formation of dense oxide film for titanium matrix composites. And with the increase of the volume fraction of reinforcements, the thickness of the oxide film decreases. As shown in figure14, comparing with TiC, reinforcement of TiB plays a much more important role in improving the antioxidant in titanium matrix composites.

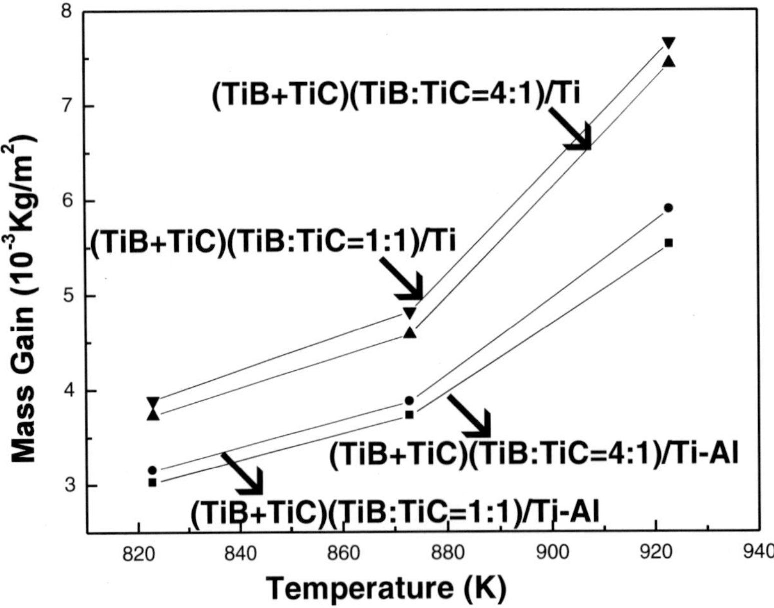

Figure 14. Mass gain-temperature relationship of titanium matrix composites.

Qin [63] has studied the oxidation behavior of Ti6242 composites reinforced with TiB and TiC. In the outer layer of oxide, continuous Al_2O_3-based TiO_2 and Al_2O_3 mixed oxide layer is observed, followed by Ti-rich mixture of TiO_2 and Al_2O_3 oxide layer, and then the Al-rich mixture of TiO_2 and Al_2O_3 oxide layer, which is also the oxygen diffusion layer. The aluminum accelerates the appearance of equiaxed and approximate equiaxed TiC reinforcement, and the appearance of smaller TiB particle reinforcement contributes much to the formation of denser oxide layer.

Yang [64] has made some researches on the high-temperature oxidation behavior of (TiC + TiB + Nd_2O_3) / Ti composites, whose antioxidant is higher than that of TiC / Ti composites. In multi-reinforced titanium matrix composites, with the increase of the volume fraction of Nd_2O_3 reinforcements, the activation energy of oxidation increases, the thickness of oxide decreases and antioxidant increases. The addition of Al_2Nd refines matrix and makes the oxide layer denser, which can improve the high temperature oxidation resistance for the multi-reinforced titanium matrix composites.

8. Superplastic Deformation Behavior

Wang [65-66] studied the deformation behavior of in situ TiB + TiC / Ti composites. The superplasticity of 5vol.% TiB and TiC reinforced titanium matrix composites material is better than that of titanium matrix composites reinforced with 1vol.% or 10vol.% TiB and TiC. The largest elongation of about 659% was obtained, as shown in figure 15. The addition of reinforcements can refined grain, pin the dislocation and grain boundaries, promote the transmission of flow stress, reduce the superplastic temperature and therefore improve the superplasticity of composites.

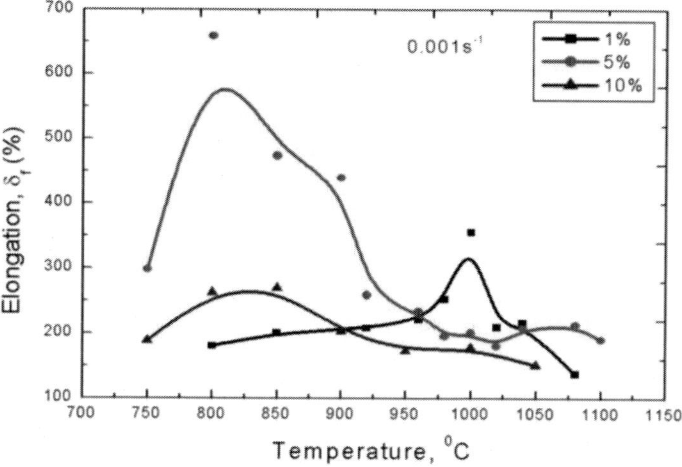

Figure 15. Effect of test temperature on elongation of TMCs.

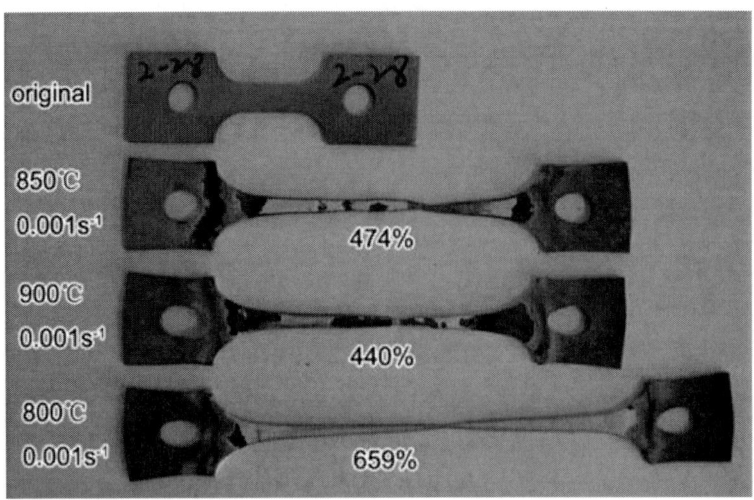

Figure 16. Profiles of the deformed specimens.

Figure 16 shows the profiles of the superplastic-deformed titanium matrix composites. After superplastic deformation, no obvious necking is observed, which means that uniform deformation takes place even when the elongation is 659%.

Ma [67-71] has reported the thermal processing and thermal deformation of (TiB + TiC)/Ti-1100 composites and TiC/Ti-1100 composites. The effect of forging process on the room temperature and high temperature tensile properties of in situ titanium matrix composites is described as follows: Widmanstatten is obtained in β forging, which has high strength and low plasticity. Duplex microstructure is obtained in conventional forging, which has good ductility and low strength. After the composites forged in high-temperature in α zone, equiaxed α phase is obtained, which has high strength and plasticity. Figure 17 shows the optical micrographs of *TMCs* deformed at different temperature. Table 3 shows the effect of forging temperature on and mechanical properties of TMCs.

Table 3. Effect of forging temperature on and mechanical properties of TMCs

Thermomechanical processing temperatures (℃)	E (GPa)	Y.S (MPa)	U.T.S (MPa)	Elongation (%)
1150	127.7	1167.4	1182.7	5.3
1060	129.3	1178.3	1193.2	6.8
900	130.2	1355.7	1378.6	6.3

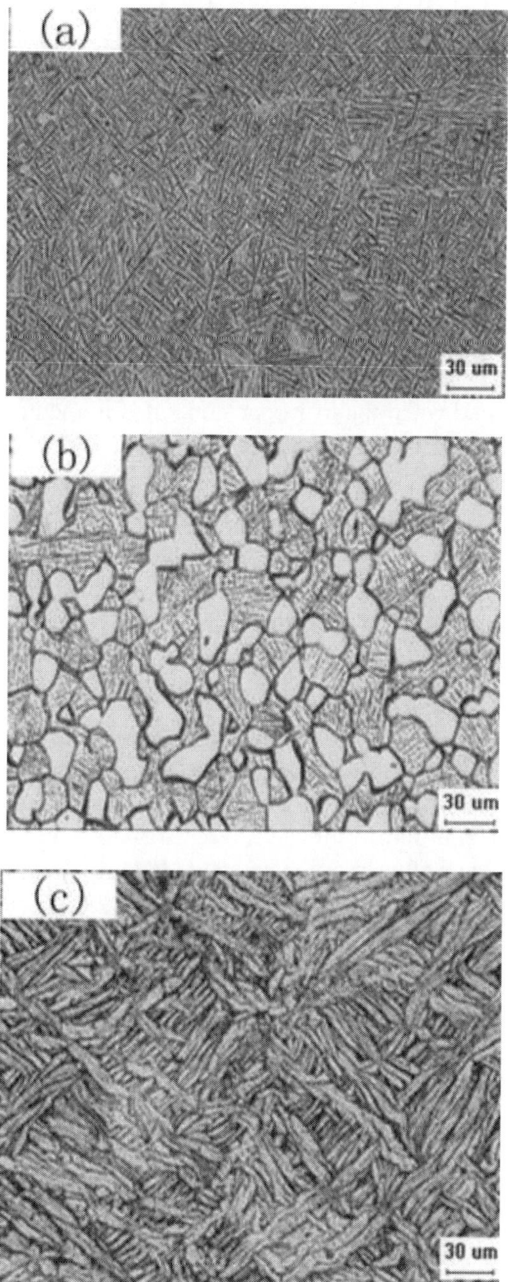

Figure 17. Optical micrographs of TMCs deformed at different temperature: (a)1150℃, (b)1060℃, (c)900℃.

9. Hydrogenated Processing and Heat Treatment

The thermal processing behavior of hydrogenated (TiB + TiC)/Ti-6Al-4V composites has been studied by Lu [72]. The flow stress of hydrogenated composites decreases with the increase of temperature. In addition, flow stress is very sensitive to strain rate and with the increase of strain rate, the flow stress increases significantly. The addition of hydrogen lowers activation energy of the composites. When the deformation temperature is higher than 825 °C, at the fixed hydrogen content the phase transition temperature is equal to the corresponding deformation temperature, which is just the optimal hydrogen content. When the deformation temperature is below 825 °C, the optimal hydrogen content is 0.40 wt.%. Figure 18 shows the optimum hydrogen concentration (OHC) at different deformation temperature.

Lu has also studied the effect of hydrogen on the superplastic deformation behavior of the TiB + TiC mixed strengthened Ti-6Al-4V composites [73]. It can be found that the addition of hydrogen decreases the super-plastic flow stress of the matrix alloy and composite materials. With the increase of hydrogen content, the flow stress decreases and then increases. The lowest value is obtained when the hydrogen content is in medium level, which decreases by more than 50%. When the hydrogen content is lower, β-phase volume fraction increases and the flow stress decreases. Optical micrographs of as-received (0 wt.% H) and hydrogenated composite are shown in Fig 19. The matrix of as-received composite was composed by equiaxed or near-equiaxed α phase and transformed β microstructure with lamellar β structures. The needle-shaped whisker was TiB, while the equiaxed particle was TiC. The volume fraction of β phase significantly increases with increasing hydrogen concentration. At 0.60 wt.% H, almost all lamellar α phase transformed to β phase (Fig. 19(d)).

Li has studied the effect of heat treatment on thermal stability of Ti matrix composite mechanical properties of TMCs [74]. Compared with TMCs without thermal exposure, ultimate strength of TMCs after thermal exposure increases slightly, the ductility sharply decreases. After thermal exposure, ultimate strength of TMCs treated by TRIPLEX heat treatment is higher than that of β heat treatment. The ductility of TMCs treated by TRIPLEX heat treatment is better than that of β heat treatment. Figure 20 shows the preparing process of large-size TMCs.

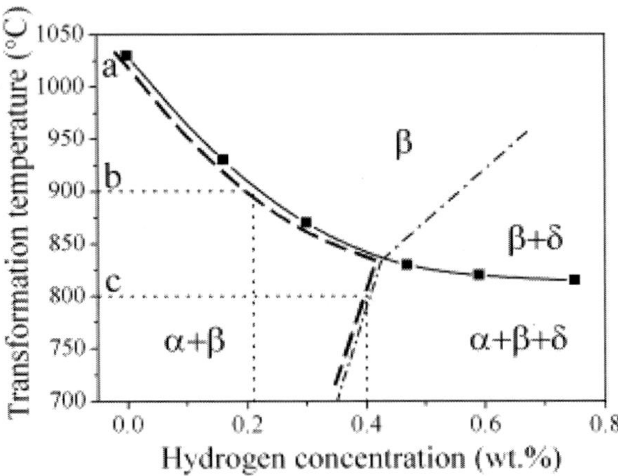

Figure 18. Optimum hydrogen concentration (OHC) at different deformation temperature.

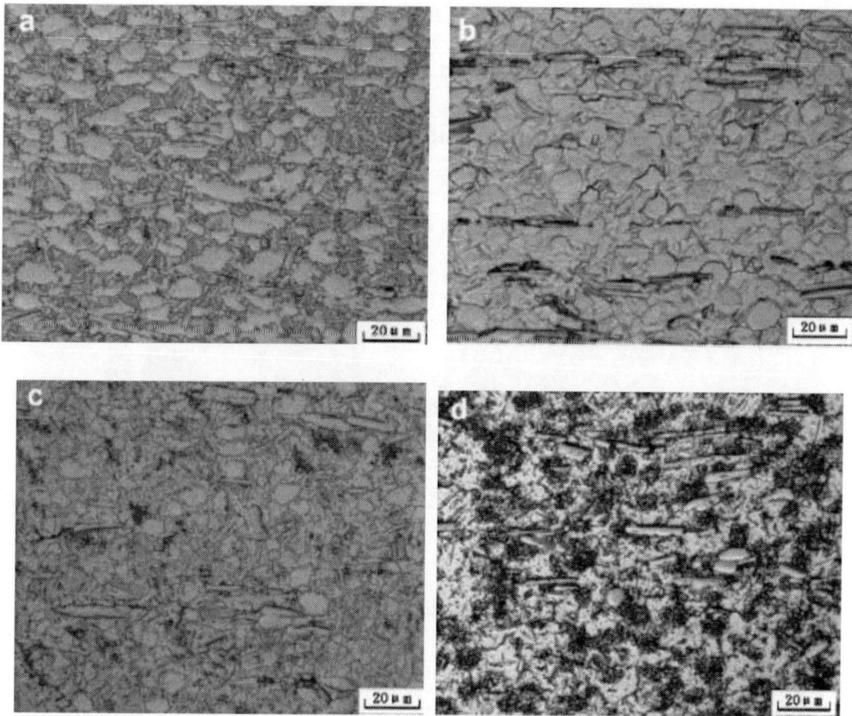

Figure 19. Optical micrographs of as-received and hydrogenated composites: (a) 0 wt.% H, (b) 0.15 wt.% H, (c) 0.35 wt.% H, and (d) 0.60 wt.% H.

Figure 20.(continued)

Figure 20. Preparing process of large-size TMCs.

CONCLUSION

In summary, the processing technology of in situ titanium matrix composites is similar to that of titanium alloys. Comparing with titanium alloys, with the similar cost, the mechanical properties and especially high-temperature mechanical properties of in situ titanium matrix composites are better, which is very important for the applications in military and civilian structures. However, there are still some questions that should be solved quickly, for example, how to produce titanium matrix composites with low-cost, in which high-performance efficient technology is a key point.

1) A new in situ synthesis technology which is low-cost and high-efficient has been developed to prepare the titanium matrix composites. The reinforcements with different shapes and sizes could be synthesized utilizing the chemical reaction between titanium and B_4C, C or ReB_6.
2) The effects of the distribution, size, content and microstructure of reinforcements on the mechanical and physical properties of titanium matrix composites have been carried out. Owing to the addition of reinforcements, some mechanical and physical properties of titanium matrix are improved to some extent.
3) In addition, the effects of hydrogen on the thermal processing and the superplastic deformation behaviors of in situ titanium matrix composites have been also investigated. And the proper addition of hydrogen could lower activation energy and decrease the flow stress at high temperature of titanium matrix composites, which is beneficial to reduce the producing cost. Llarge-size TMCs about 1500kg is also prepared by in situ synthesis technology.

REFERENCES

[1] Abkowitz S, Abkowitz SM, Fisher H, et al. *CermeTi® discontinuously reinforced Ti-matrix composites: manufacturing, properties and applications*. JOM, 2004, 56: 37-41.

[2] Saito T: *Automotive application of discontinuously reinforced TiB-Ti composites*. JOM, 2004, 56: 33-36.

[3] Tjong SC, Mai YW: *Processing-structure-property aspects of particulate- and whisker-reinforced titanium matrix composites.* Composites Science and Technology, 2008, 68: 583-601.

[4] Tjong SC, Ma ZY: *Microstructural and mechanical characteristics of in situ metal matrix composites.* Materials Science and Engineering R, 2000, 29: 49-113.

[5] Ranganath S: *A review on particulate-reinforced titanium matrix composites.* Journal of MaterialsScience, 1997, 32: 1-6.

[6] Alman DE, Hawk JA: *The abrasive wear of sintered titanium matrix–ceramic particle reinforced composites*. Wear, 1999, 225-229: 629-639.

[7] Abkowitz S, Weihrauch PF, Abkowitz FH, Heussi HL. *The commercial application of low-cost titanium composites*. JOM, 1995, 47: 40-41.

[8] Loretto MH, Konitzer DG: *The effect of matrix reinforcement reaction on fracture in Ti–6Al–4V-base composites*. Metallurgical Transactions A, 1990, 21: 1579-1587.

[9] Choi SK, Chandrasekaran M, Brabers MJ: *Interaction between titanium and SiC.* Journal of Materials Science, 1990, 25: 1957-1964.

[10] Lu YC, Sass SL, Bai Q, et al. *The influence of interfacial reactions on the fracture toughness of Ti-Al2O3 interfaces*. Acta Metallurgica et Materialia, 1995, 43: 31-41.

[11] Ma XY, Li CR, Zhang WJ: *The thermodynamic assessment of the Ti–Si–N system and the interfacial reaction analysis*. Journal of Alloys and Compounds, 2005, 394: 138-147.

[12] Lemus-Ruiz J, Aguilar-Reyes EA: *Mechanical properties of silicon nitride joints using a Ti-foil interlayer.* Materials Letters, 2004, 58: 2340-2344.

[13] Gotman I, Gutmanas EY: *Interaction of Si3N4 with titanium powder.* Journal of Materials Science Letters, 1990, 9: 813-815.

[14] Lu WJ, Zhang D, Zhang XN, et al. *HREM study of TiB/Ti interfaces in a TiB-TiC in situ composite.* Scripta Materialia, 2001, 44: 1069-1075.

[15] Konitzer DG, Loretto MH: *Microstructural assessment of Ti–6Al–4V–TiC metal matrix composite.* Acta Metallurgica, 1989, 37: 397-406.

[16] Yang ZF, Lu WJ, Qin JN, et al. *Microstructure and tensile properties of in situ synthesized (TiC+TiB+Nd2O3)/Ti-alloy composites at elevated temperature.* Materials Science and Engineering A, 2006, 425: 185-191.

[17] Sastry SML, Peng TC, Beckerman LP: *Structure and properties of rapidly solidified dispersion-strengthened titanium alloys: II. Tensile and creep properties.* Metallurgical Transactions A, 1984, 15: 1465-1474.

[18] Xiao L, Lu WJ, Li YG, et al. *Thermal stability ofin situ synthesized high temperature titanium matrix composites*. Journal of Alloys and Compounds, 2009, 467: 135-141.

[19] Gorsse S, Chaminade JP, Petitcorps YL : *In situ preparation of titanium base composites reinforced by TiB single crystals using a powder metallurgy technique.* Composites A, 1998, 29: 1229-1234.

[20] Chandran KSR, Panda KB, Sahay SS: *TiBw-reinforced Ti composites: processing, properties, application prospects and research needs*. JOM, 2004, 56: 42-48.

[21] Wang L, Niinomi M, Takahashi S, et al. *Relation between fracture toughness and microstructure of Ti-6Al-2Sn-4Zr-2Mo alloy reinforced with TiB particles.* Materials Science and Engineering A, 1999, 263: 319-325.

[22] Kobayashi M, Funami M, Suzuki S, et al. *Manufacturing process and mechanical properties of fine TiB dispersed Ti–6Al–4V alloy composites obtained by reaction sintering.* Materials Science and Engineering A, 1998, 243: 279-284.

[23] Bhat BVR, Subramanyam J, Bhanu Prasad VV: *Preparation of Ti–TiB–TiC & Ti–TiB composites by in situ reaction hot pressing*. Materials Science and Engineering A, 2002, 325: 126-130.

[24] Panda KB, Chandra KSR: *Synthesis of ductile titanium-titanium boride (Ti-TiB) composites with a beta titanium matrix: the nature of TiB formation and composite properties.* Metallurgical and Materials Transactions A, 2003, 34: 1371-1385.

[25] Suryanarayana S: *Mechanical alloying and milling*. Progress in Materials Science, 2001, 46: 1-184.

[26] Godfrey TM, Wisbey A, Goodwin PS, et al. *Microstructure and tensile properties of mechanically alloyed Ti-6Al-4V alloy with B additions.* Materials Science and Engineering A, 2000, 282: 240-250.

[27] Yamamoto T, Otsuki A, Ishihara K, et al. *Synthesis of near net shape high density TiB/Ti composite.* Materials Science and Engineering A, 1997, 239-240: 647-651.

[28] Nakane S, Yamada O, Miyamoto Y, et al. *Simultaneous synthesis and densification of TiB/α-Ti(N) composite material by self-propagating combustion under nitrogen pressure.* Solid State Communications, 1999, 110: 447-450.

[29] Fu ZY, Wang H, Wang WM, et al. *Composites fabricated by self-propagating high-temperature synthesis.* Journal of Materials Processing Technology, 2003, 137: 30-34.

[30] Zhang E, Zeng S, Zhu Z: *Microstructure of XDTM Ti-6Al/TiC composites.* Journal of Materials Science, 2000, 35: 5989-5994.

[31] Fan Z, Miodownik AP: *Microstructural evolution in rapidly solidified Ti-7.5Mn-0.5B alloy*. Acta Materialia, 1996, 44: 93-110.

[32] Rangarajan S, Aswath PB, Soboyejo WO: *Mirostructure development and fracture of in-situ reinforced Ti-8.5Al-1B-1Si*. Scripta Materialia, 1996, 35: 239-245.

[33] Ranganath S, Vijayakumar M, Subrahmanyan J: *Combustionassisted synthesis of Ti-TiB-TiC composite via the casting route*. Materials Science and Engineering A, 1992, 149: 253-357.

[34] Banerjee R, Genc A, Collins PC, et al. *Comparison of microstructural evolution in laser-deposited and arc-melted in situ Ti-TiB composites.* Metallurgical and Materials Transactions A, 2004, 35: 2143-2152.

[35] Srivatsan TS, Soboyejo WO, Lederich RJ: *Tensile deformation and fracture behavior of a titanium-alloy metal-matrix composite.* Composites A, 1997, 28: 365-376.

[36] Li BS, Shang JL, Guo JJ, et al. *In situ observation of fracture behavior of in situ TiBw/Ti composites.* Materials Science and Engineering A, 2004, 383: 316-322.

[37] Zhang XN,Lu WJ, Zhang D, et al. *In situ technique for synthesizing (TiB+TiC)/Ti composites.* Scripta Materialia, 1999, 44: 39-46.

[38] W. J. Lu, D. Zhang, X.N. Zhang, R.J. Wu. *Microstructure and Mechanical Properties of In Situ Synthesized TiC/Ti Composites*. Journal of Shanghai Jiaotong University, 2001(35):643-650.

[39] W. J. Lu, D. Zhang, X.N. Zhang, R.J. Wu. *Microstructure and Mechanical Properties of in situ Synthesized TiB/Ti Composites*. Journal of Shanghai Jiaotong University, 2000(34):1606-1614.

[40] Yang ZF, Lu WJ, Xu D, et al. *In situ synthesis of hybrid and multiple-dimensioned titanium matrix composites*. Journal of Alloys and Compounds, 2006, 419: 76-80.

[41] W. J. Lu, D. Zhang, X.N. Zhang, R.J. Wu. *Growth Mechhanism of Reinforcement in IN SITU Processed TiC/Ti Composites*. Acta Metallrugica Sinica, 1999(35): 536-540.

[42] W. J. Lu, D. Zhang, X.N. Zhang, R.J. Wu. *Growth Mechhanism of Reinforcement in IN SITU Processed TiB/Ti Composites* Acta Metallrugica Sinica, 2000(36): 104-108.

[43] de Graef M, Loefvander JPA, mc Cullough C, et al. *The evolution of metastable Bf borides in a Ti-Al-B alloy*. Acta Metallurgica et Materialia, 1992, 40: 3395-3406.

[44] H. Zhang, W.L. Gao, E.L. Zhang, Y.X. Jin, S.Y. Zeng. *Growth Mechanism of Primary Hollow TiB in Ti-A1-B Alloy*. Acta Materiae Compositae Sinica, 2001(18):50-53.

[45] Geng K, Lu W, Zhang D: *In situ synthesized (TiB+Y2O3)/Ti composites*. Journal of Materials Science Letters, 2003, 22: 877-879.

[46] Geng K, Lu W, Yang Z, et al. *In situ preparation of titanium matrix composites reinforced by TiB and Nd2O3*. Materials Letters, 2003, 57: 4054-4057.

[47] Yang ZF, Lu WJ, Qin JN, et al. *Microstructural characterization of Nd2O3 in in situ synthesized multiple-reinforced (TiB + TiC +Nd2O3)/Ti composites*. Journal of Alloys and Compounds, 2006, 425: 379-383.

[48] Lu WJ, Xiao L, Xu D, et al. *Microstructural characterization of Y2O3 in in situ synthesized titanium matrix composites*. Journal of Alloys and Compounds, 2007, 433: 140-146.

[49] Zhang D, Geng K, Qin Y, et al. *The orientation relationship between TiB and RE2O3 in in situ synthesized titanium matrix composites*. Journal of Alloys and Compounds, 2005, 392: 282-284.

[50] Lu WJ, Zhang D, Zhang XN, et al. *Microstructure and tensile properties of in situ (TiB+TiC)/Ti6242 (TiB:TiC=1:1) composites prepared by common casting technique*. Materials Science and Engineering A, 2001, 311: 142-150.

[51] Yang Z, Lu W, Zhao L, et al. *Microstructure and mechanical property of in situ synthesized multiple-reinforced (TiB + TiC + La2O3) / Ti composites*. Journal of Alloys and Compounds, 2008, 455: 210-214.

[52] Geng K, Lu WJ, Zhang D: *Microstructure and tensile properties of in situ synthesized (TiB + Y2O3) / Ti composites at elevated temperature*. Materials Science and Engineering A, 2003, 360: 176-182.

[53] Yang ZF, Lu WJ, Qin JN, et al. *Microstructure and tensile properties of in situ synthesized (TiC+TiB+Nd2O3)/Ti-alloy composites at elevated temperature*. Materials Science and Engineering A, 2006, 425: 185-191.

[54] Xiao L, Lu W, Yang Z, et al. *Effect of reinforcements on high temperature mechanical properties of in situ synthesized titanium matrix composites*. Materials Science and Engineering A, 2008, 491: 192-198.

[55] Xiao L, Lu W, Qin J, et al. *High-temperature tensile properties of in-situ synthesized titanium matrix composites with strong dependence on strain rates.* Journal of Materials Research, 2008, 23: 3066-3074.

[56] Lu W, Zhang D, Zhang X, et al. *Creep rupture life of in situ synthesized (TiB+TiC)/Ti matrix composites.* Scripta Materialia, 2001, 44: 2449-2455.

[57] Xiao L, Lu WJ, Qin J, et al. *Creep behaviors and stress regions of hybrid reinforced high temperature titanium matrix composite.* Composites Science and Technology, 2009, 69: 1925-1931.

[58] Xiao L, Lu WJ, Li YG, et al. *Steady state creep of in-situ TiB plus La2O3 reinforced high temperature titanium matrix composite.* Materials Science and Engineering A. 2009 499 (1-2): 500-506.

[59] Qin Y, Lu W, Qin J, et al. *Oxidation Behavior of In Situ Synthesized TiB/Ti Composite in Air Environment.* Materials Transactions, 2004, 45: 3241-3246.

[60] Qin Y, Zhang D, Lu W, et al. *Oxidation Behavior of In Situ Synthesized TiB/Ti-Al Composite.* Journal of Materials Science, 2005, 40: 6553-6558.

[61] Qin Y, Lu W, Zhang D, et al. *Oxidation of In Situ Synthesized TiC Particle-Reinforced Titanium Matrix Composites.* Materials Science and Engineering A, 2005, 404: 42-48.

[62] Qin Y, Zhang D, Lu W, et al. *Oxidation behavior of in situ synthesized (TiB + TiC)/Ti–Al composites.* Materials Letters, 2006, 60: 2339-2345.

[63] Qin Y, Zhang D, Lu W, et al. *A new high-temperature, oxidation-resistant in situ TiB and TiC reinforced Ti6242 alloy.* Journal of Alloys and Compounds, 2008, 455: 339-375.

[64] Yang ZF, Lu WJ, Qin JN, et al. *Oxidation behavior of in situ synthesized (TiC + TiB + Nd2O3)/Ti composites.* Materials Science and Engineering A, 2008, 472: 187-192.

[65] Wang MM, Lu WJ, Qin JN, et al. *The effect of reinforcements on superplasticity of in situ synthesized (TiB+TiC)/Ti matrix composite.* Scripta Materialia, 2006, 54: 1955-1959.

[66] Wang MM, Lu WJ, Qin JN, et al. *Superplastic behavior of in situ synthesized (TiB + TiC)/Ti matrix composite.* Scripta Materialia, 2005, 53: 265-270.

[67] Ma F, Lu W, Qin J, et al. *Hot deformation behavior of in situ synthesized Ti-1100 composite reinforced with 5 vol.% TiC particles.* Materials Letters, 2006, 60: 400-405.

[68] Ma F, Lu W, Qin J, et al. *The effect of forging on microstructure and mechanical properties of in situ TiC/Ti composites.* Materials Transactions, 2006, 47: 1322-1327.

[69] Ma F, Lu W, Qin J, et al. *Effect of Forging Temperature on Microstructure and Mechanical Properties of In Situ (TiB+TiC)/Ti Composites.* Materials Transactions, 2006, 47: 1750-1754.

[70] Ma F, Lu W, Qin J, et al. *Effect of forging and heat treatment on the microstructure of in situ TiC/Ti-1100 composites.* Journal of Alloys and Compounds, 2007, 428: 332-337.

[71] Ma F, Lu W, Qin J, et al. *The effect of forging on mechanical properties of (TiB+TiC)/Ti-1100 composites at elevated temperature.* Materials & Design, 2007, 28: 1339-1342.

[72] Lu J, Qin J, Lu w, et al. *Effect of hydrogen on microstructure and high temperature deformation of (TiB + TiC) Ti-6Al-4V composite.* Materials Science and Engineering A, 2009, 500: 1-7.

[73] Lu J, Qin J, Lu w, et al. *Effect of hydrogen on superplastic deformation of (TiB+TiC)/Ti–6Al–4V composite*, International Journal of Hydrogen Energy, 2009, doi:10.1016/j.ijhydene.2009.07.091

[74] Jiuxiao Li, Liqiang Wang, Jining Qin, Yifei Chen, Weijie Lu, Di Zhang. *The effect of heat treatment on thermal stability of Ti matrix composite.* Journal of Alloys and Compounds 509, 2011: 52–56.

In: Metal Matrix Composites
Editor: J. Paulo Davim

ISBN: 978-1-61209-771-8

Chapter 6

IMPACT BEHAVIOUR OF SIMILAR AND DISSIMILAR LINEAR FRICTION WELDING JOINTS BETWEEN A 2024 AL ALLOY AND A 2124-25VOL.%SICP COMPOSITE

L. Ceschini[1], A. Morri[1], F. Rotundo[*2], G. L. Garagnani[3] and M. Merlin[3]

[1]SMETEC Dept., University of Bologna, V. le Risorgimento, Bologna, Italy

[2]DIEM, University of Bologna, V.le Risorgimento, Bologna, Italy

[3]Engineering Department, University of Ferrara, Via Saragat, Ferrara, Italy

ABSTRACT

This chapter analyzes the effects of Linear Friction Welding (LFW) on the impact behaviour of an aluminium matrix composite (AA2124/25%vol.SiCp) and an unreinforced Al alloy (AA2024). The aluminum based metal matrix composite (MMC) was obtained by powder metallurgy, forged and T4 heat-treated, while the Al alloy was supplied in the extruded and T4-heat treated condition. Optical Microscopy (OM) and Scanning Electron Microscopy (SEM) with Energy Dispersive Spectroscopy (EDS) were used to characterize the effects of the welding process on the microstructural characteristics of the LFW joints, which were obtained between similar (MMC-MMC, AA2024-AA2024) and dissimilar materials (MMC/AA2024). Instrumented impact tests were carried out on Charpy specimens, machined either from the base materials or from the LFW joints. The mechanisms of failure was investigated by SEM analyses of the fracture surfaces. The microstructure of the joints was found to be dependent on both the initial microstructure of the tested materials and the welding process. The LFW induced a reduction of the impact strength of the AA2024 alloy, while it had no relevant effects on the impact strength of the AA2124/25%SiCp composite. The impact energy of the dissimilar weld was comparable to that of the MMC joint, since fracture propagated

[*]Corresponding author, E-mail: fabio.rotundo@gmail.com

mostly on the composite side. The fracture path was usually located along the thermo-mechanically affected zone in the similar joints.

1. INTRODUCTION

Silicon carbide particle (SiCp) reinforced aluminium-based composites are the most common and commercially available metal matrix composites (MMCs), with applications in aerospace, motorsport and auto manufacturing industries, due to their excellent specific mechanical properties and to their relatively easy production route [1-4]. One of the main advantages of particulate reinforced MMCs, compared with long-fibre reinforced ones, is that they can be processed by the same technologies developed for corresponding monolithic alloys (such as extrusion, forging, hot and cold rolling); moreover, they possess nearly isotropic mechanical properties [4-6]. However, the industrial application of Al-based MMCs has hitherto been limited, mainly due to their low weldability, since traditional fusion welding processes generally lead to microstructural defects, which result in a general decrease in the mechanical properties, such as strength, ductility and impact behaviour [7-8]. In particular, SiCp reinforced Al-based MMCs have been welded by brazing [9], laser welding [10] and arc fusion welding (tungsten inert gas (TIG) and metal inert gas (MIG)) [11-12]. Nevertheless, besides the typical defects commonly found in Al alloy fusion welds (solidification shrinkages, oxide inclusions, gas pores, etc.), the addition of the ceramic reinforcement also induces: higher viscosity of MMCs melts; particle segregation and clustering; undesired matrix-reinforcement reactions, with the formation of coarse and brittle phases (e.g. A14C3,) and the evolution of occluded gases during welding, thus limiting the achievement of sound joints [8].

In solid state joining processes, on the other hand, joining is obtained at temperatures substantially lower than the melting point of the base material, without the addition of brazing filler metal, producing higher quality joints and also allowing the welding of dissimilar metals [13-14]. In friction welding (FW), joining is obtained by frictional heating, produced from the mechanically-induced sliding motion between rubbing surfaces, held together under pressure. Once sufficient heat has been generated, the weld is consolidated in a forging stage, with increased pressure. Resulting joints are usually characterised by a narrow heat-affected zone and a thermo-mechanically affected zone, which undergoes severe plastic deformation due to the high local temperature and internal pressures. The main advantages of friction welding are the high repeatability and excellent properties of the weld. Moreover, no special preparation of the sample is needed, no filler wire or shielding gas is required (despite the affinity of Al with oxygen) and no, or little, pollution and waste are produced during the welding process. Friction welding techniques include: friction stir welding (FSW), inertia or rotary friction welding (RFW) and linear friction welding (LFW).

In Friction Stir Welding (FSW) a cylindrical rotating tool, supplied with a pin of smaller diameter extending from the tool shoulder, is moved at the interface between two components, generating frictional heating and severe plastic deformation of the material, due to the stirring effect of the pin. Although this process was initially developed to weld aluminium alloys [15-17], several studies demonstrated that it can also be successfully applied to weld particle reinforced Al based composites [18-23]. One of the main limitations of this technique, when applied to particle reinforced MMCs, can be the severe wear of the tool, due to the abrasive action of the ceramic reinforcement [18]. This abrasion could be limited by means of recently developed highly-wear resistant FSW tools [24], or avoided by the use of other friction welding techniques, such as Linear Friction Welding (LFW), which widen the application of rotary friction welding techniques to non axisymmetric components.

In LFW, one component is supplied with a linear oscillating motion and brought in contact with the other component, which is rigidly clamped. The concurrent action of an increasing axial force (resulting in increasing pressure at the interface) and the reciprocating motion, supply the frictional heating necessary for the welding process. The true contact area increases and, once sufficient surface plasticization has occurred, the deformed material is expelled as flash, leading to an axial shortening which increases in an approximately linearly manner with time. The reciprocating motion is stopped as soon as the imposed axial shortening (burn-off distance) is achieved, and a forging force is maintained to consolidate the weld, involving further axial shortening.

LFW was successfully applied to Ti alloys [25-27], steels [28] and Ni-based superalloys [29], and only recently its application to Al-based MMCs was studied [30-31], without addressing its impact behaviour, which is a fundamental property for the evaluation of the weld properties. The present chapter is aimed at illustrating the effects of Linear Friction Welding (LFW) on the impact behaviour of the AA2124/25%vol.SiCp composite, the AA2024 alloy and dissimilar joints between the composite and the unreinforced alloy, correlating the impact test results with the microstructural modifications induced by the process.

2. Experimental

The composite material investigated in this study is the AMC225xe Al based MMC, which was obtained by powder metallurgy combining a AA2124 Al matrix (Table 1) with 25vol.% of fine SiC particles, with a nominal size of 3 μm. In this proprietary production route, fine aluminium and SiC powders were first mixed and degassed, then MMC billets were produced by hot isostatic pressing. 15 mm thick plates were obtained by forging (with a deformation ratio of about 6:1) and then heat treated at the T4 condition. A AA2024 alloy was selected to also study the effects of LFW on the unreinforced alloy. This Al alloy, supplied as extruded rectangular plates and T4 heat treated, has the same nominal chemical composition (Table 1) as the MMC matrix, except for a slight difference in the Fe and Si content.

Specimens, with a 15x36 mm^2 cross section, were Linear Friction Welded at TWI (The Welding Institute, Cambridge, UK), according to the scheme shown in Fig.1. Welding parameters were optimized after some preliminary tests and are reported in Table 2.

Table 1. Nominal chemical compositions (wt.%) of the AA2124 MMC matrix and unreinforced AA2024

	Cr	Cu	Fe	Mg	Mn	Si	Ti	Zn	others	Al
AA2124	≤0.10	3.80-4.90	≤0.30	1.20-1.80	0.30-0.90	≤0.20	≤0.15	≤0.25	≤0.15	Balance
AA2024	≤0.10	3.80-4.90	≤0.50	1.20-1.80	0.30-0.90	≤0.50	≤0.15	≤0.25	≤0.15	Balance

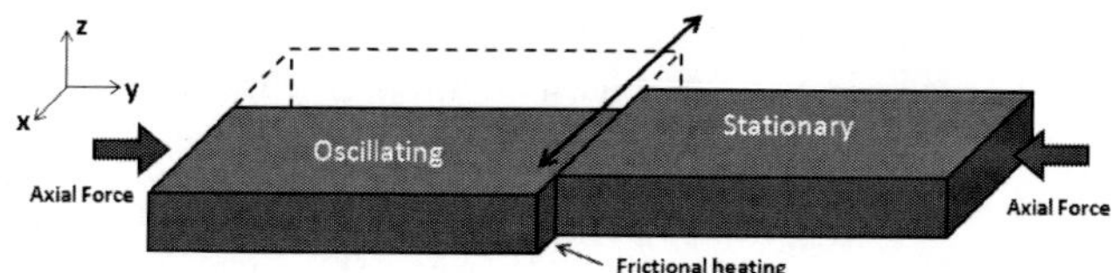

Figure 1. Linear friction welding schematic.

Table 2. Linear Friction Welding parameters for similar and dissimilar joints

	Friction/Forge Force [kN]	Pressure [N/mm^2]	Frequency [Hz]	Set Amplitude [mm]	Burn-off [mm]
MMC-MMC	100	185	50	±2	2
Al-Al	85	157	50	±2	2
Al-MMC	85	157	50	±2	2

Samples for microstructural investigations were cut along the *y-z* plane in Fig.1. The metallographic samples were mechanically ground, coarse polished (according to ASTM E3 [32]) and chemically etched with Keller's reagent. The microstructural characterization was carried out by optical microscopy (OM) under polarized light and scanning electron microscopy (SEM) with an energy dispersive spectroscopy (EDS) microprobe.

Standard Charpy specimens (10x10x55 mm^3) were machined with a 45° V-notch of 2 mm in depth, according to ASTM E23 [33]. The notch, with a final radius of 0.25 mm, was machined in the weld centre (Figs.2-3), parallel to the *x*-axis.

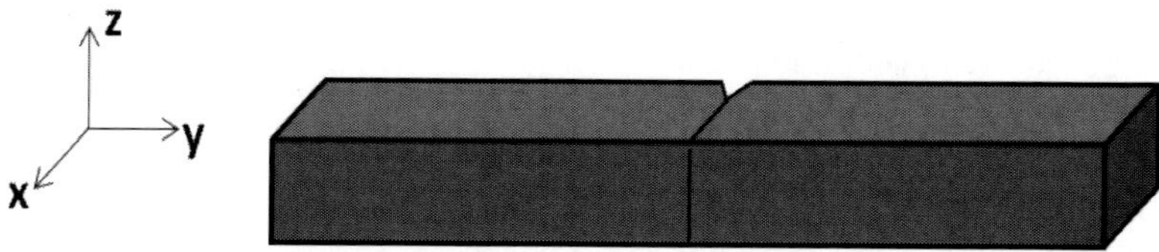

Figure 2. Charpy sample orientation with respect to the welding line.

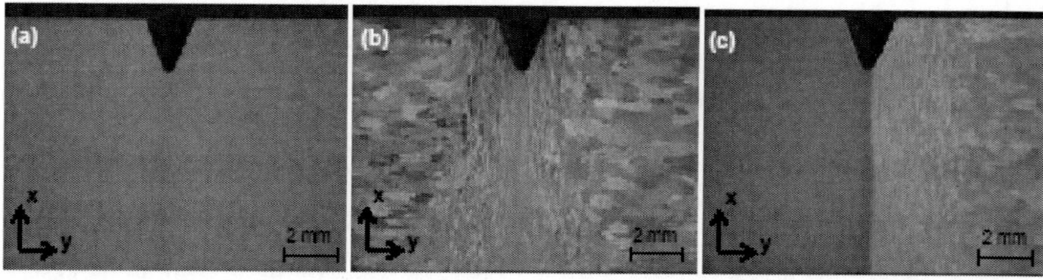

Figure 3. Etched low magnification micrographs of the V notch in the Charpy specimens for: (a) MMC-MMC LFW (AA2124/25%vol.SiC$_p$); (b) Al-Al LFW (AA2024); (c) dissimilar Al-MMC LFW.

Impact tests were performed at room temperature using a Ceast instrumented pendulum with an available energy of 50 J. The machine is equipped with an auto-calibration system of the hammer, in order to execute the adjustments due to pendulum frictions and air resistance, and with a DAS8000 data acquisition system. Impact properties were measured according to UNI EN ISO 14556:2003 [34]. Total absorbed energy (W_t) was calculated as the integral of the load-displacement curve with the end assumed at 2% of its peak. The two complementary contributions of the crack nucleation energy (W_m) and crack propagation energy (W_p) were evaluated, where W_m is the energy absorbed until the maximum load (F_m) is achieved, and W_p is the energy absorbed from F_m to the end of the test. A total number of 4 specimens for each similar joint type and base materials were tested, while 6 specimens were used for the dissimilar MMC/AA2024 joint.

An additional criterion to evaluate ductility in impact tests is the measurement of lateral expansion, which was carried out according to ASTM E23, i.e. measuring the expansion relative to the plane defined by the undeformed specimen portion, on both sides of each specimen half. Finally, fracture surfaces were investigated by means of stereomicroscope and SEM, and metallographic cross-sections were observed by optical microscopy, in order to evaluate the influence of the LFW process on the mechanisms of failure.

3. Results and Discussion

MMCs produced by the powder metallurgy process are generally characterized by a very fine microstructure, which means that optical microstructural analysis is generally unable to resolve the grain boundaries of the Al matrix. However, it made it possible to observe the particles reinforcing distribution and, in particular, the presence of large particle-free bands, which were caused by the forging process in the studied AA2124/25%vol.SiC_p composite (Fig.4-a).

The AA2024 alloy, instead, was characterized by the presence of a fibrous grain structure, with elongated grains that are about 800 μm long in the extrusion direction, about 300 μm in the thickness direction (Fig.4-b).

Linear friction joints were characterized by the presence of flash material extruded in both a parallel and normal manner, with respect to the force-motion plane (*x-y)*, as shown for the MMC joint in Fig.5-a.

Etched cross-sections of the MMC LFW joints (shown in the optical micrographs in Fig.5-b and -c) emphasized the presence of three zones, both in similar and dissimilar joints. In the *Weld Centre,* a relevant grain refinement occurred, due to the concurrent effect of frictional heating and severe plastic deformation caused by the solid state welding process. The matrix grain size was probably reduced on a sub-micrometric scale, so that OM analyses do not allow the grain boundaries to be clearly resolved. In the case of the MMC, a very uniform particle distribution also characterizes this zone which is, however, comparable with that in the base material (Fig.6). In the *Thermo-Mechanically Affected Zone* (TMAZ), instead, the fibrosity of the material follows the plastic flow induced by the welding process (Fig.5-c).

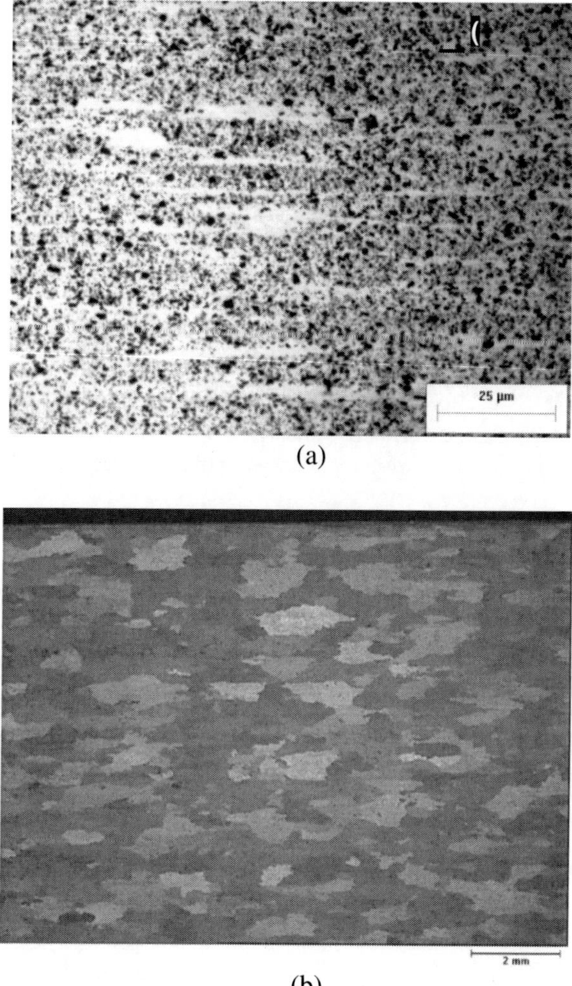

(a)

(b)

Figure 4. Microstructure of (a) the base AA2124/25%vol.SiC$_p$ composite and (b) the AA2024 Al alloy.

This fibrosity can change in the various impact specimens extracted from different sections of the joint. No plastic deformation occurred in the *Heat Affected Zone* (HAZ), but the thermal cycling probably affected the material properties. As reported in [31], in fact, in the HAZ, the LFW could induce a complete or partial dissolution of the Cu-Mg co-clusters formed during the T4 heat treatment, as well as the precipitation or coarsening of the stable S phase, with a consequent decrease in the material strength. The high temperature reached in the TMAZ leads, instead, to the formation of coarser S precipitates, while the temperature in the weld centre dissolves both the Cu-Mg co-clusters and the S precipitates and, consequently, the post-weld natural aging can lead to the formation of new Cu-Mg co-clusters.

In the dissimilar weld, a clear separation between the unreinforced AA2024 alloy and the composite was found, as shown in Fig.3-c. This is basically due to the absence of a stirring tool, such as in FSW, which could blend the two materials.

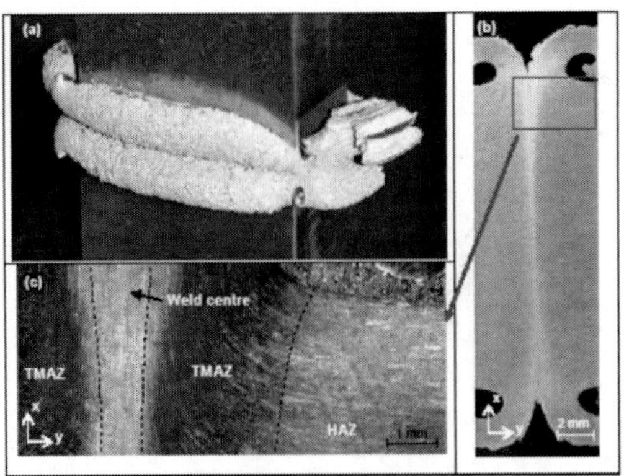

Figure 5. Side view of an AA2124/25vol.%SiC$_p$ linear friction welded joint (a); optical micrographs of the etched joint cross-section under (b) white light and (c) polarized light.

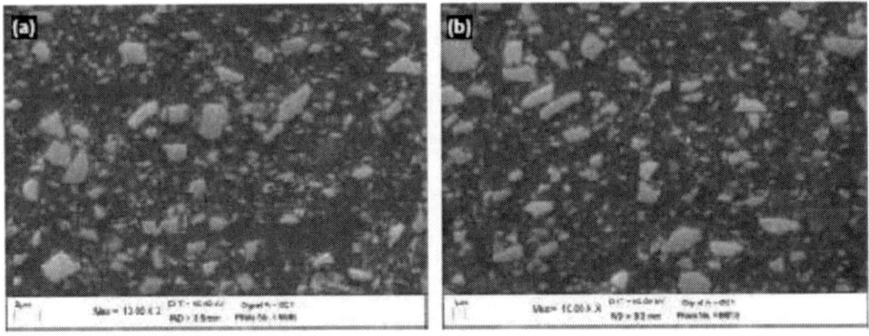

Figure 6. SEM micrographs of the (a) weld centre and (b) base composite material.

The results of the instrumented Charpy impact tests are reported in Table 3 and Fig.7.

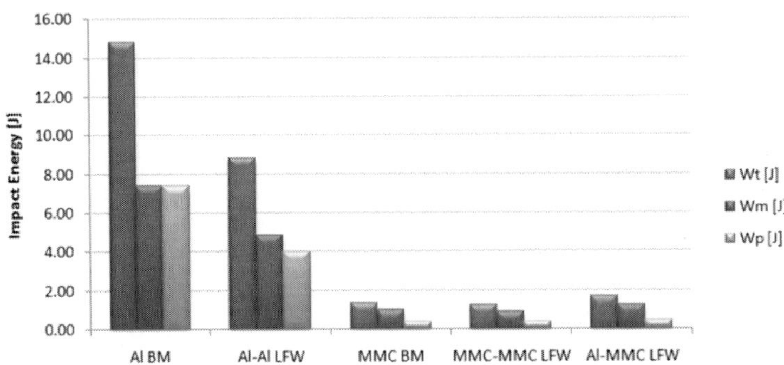

Figure 7. Comparison of the impact energies (W_t, W_m, W_p) for the AA2124/25vol.%SiC$_p$ welded (MMC-MMC LFW) and base material (MMC BM), AA2024 base (Al BM) and welded (Al-Al LFW) alloy, MMC/AA2024 dissimilar joints (Al-MMC LFW).

Table 3. Results of the Charpy impact tests performed on the: AA2124/25vol.%SiC$_p$ welded (MMC-MMC LFW) and base material (MMC BM), AA2024 welded (Al-Al LFW) and base alloy (Al BM), and on dissimilar welds (Al-MMC LFW): maximum load (F_m), total impact energy (W_t), nucleation energy (W_m) and propagation energy (W_p)

Material		Fmax [N]	Wt [J]	Wm [J]	Wp [J]
MMC-MMC LFW	1	6110	1.16	0.84	0.32
	2	7335	1.50	1.04	0.46
	3	5941	1.01	0.72	0.29
	4	6828	1.23	0.91	0.32
	Average	6553,50	1.23	0.88	0.35
	Standard dev.	647,54	0.21	0.13	0.08
MMC BM	1	7124	1.29	0.93	0.36
	2	7264	1.31	0.97	0.34
	3	7546	1.50	1.08	0.42
	4	6857	1.19	0.9	0.29
	Average	7197,75	1.32	0.97	0.35
	Standard dev.	287,06	0.13	0.08	0.05
Al-Al LFW	1	9067	8.77	5.04	3.73
	2	8729	8.08	4.56	3.52
	3	9011	9.07	5.38	3.69
	4	8926	9.34	4.41	4.93
	Average	8933,25	8.82	4.85	3.97
	Standard dev.	147,99	0.54	0.45	0.65
Al BM	1	9715	14.61	7.25	7.36
	2	9954	14.81	7.4	7.41
	3	9996	14.79	7.42	7.37
	4	10053	14.88	7.53	7.35
	Average	9929,50	14.77	7.40	7.37
	Standard dev.	148,64	0.12	0.12	0.03
Al-MMC LFW	1	5913	1.09	0.8	0.29
	2	6378	1.42	1.06	0.36
	3	6617	1.50	1.19	0.31
	4	7645	2.27	1.66	0.61
	5	7800	2.41	1.77	0.64
	6	6096	1.41	1.02	0.39
	Average	6741,50	1.68	1.25	0.43
	Standard dev.	798,43	0.53	0.38	0.15

The presence of the SiC particles reduced, by an order of magnitude, the impact strength of the MMC with respect to the unreinforced AA2024 Al alloy, with average W_t values of about 1.3±0.1 J and 14.8±0.1 J, respectively. This significant decrease in the particle reinforced MMC, with respect to the unreinforced Al alloy was already reported in literature [35].

The average total impact energy of the LFW composite (W_t=1.2±0.2 J) was, instead, substantially similar to that of the base MMC (W_t=1.3±0.1 J). This result can be clearly related to the negligible microstructural modifications (mainly in terms of defects) induced by the LFW in the composite, confirming that solid state welding processes can be more efficient that fusion welding, in the case of particle reinforced MMCs.

In the case of particle reinforced Al-based composites, produced by liquid metal processing (stir casting) and Friction Stir Welded, an increase in the impact strength of the FSW composite compared to the base material was reported in [20]. The difference in respect to the present results can be related both to the different base materials and welding processes. It is well known, in fact, that PM generally leads to finer microstructures and more homogeneous particle distribution when compared with liquid metal processes. Therefore the grain refinement was probably limited, in the LFW joints, due to the very fine microstructure of the base composite. Moreover, in the case of FSW, the presence of the pin induced a significant stirring effect, which was effective in reducing both the matrix grain size and the reinforcement particles, also inducing a more homogeneous particle distribution.

The LFW had a higher influence on the impact behaviour of the unreinforced AA2024 Al alloy, causing a decrease of about 37 % (from 14.8±0.1 J to 8.8±0.5 J) in the absorbed energy. This could be due to the elongated grain texture induced by LFW, which being parallel to the crack nucleation and propagation direction, reduced both W_m and W_p energies (Fig.7).

Moreover, the plot of Fig.7 shows that while the contributions of W_m and W_p are similar in the AA2024 alloy, in the LFW samples the nucleation energy was higher than that of the propagation. This behaviour could also be due to the fact that propagation, rather than nucleation, of a crack is mainly affected by the texture, because the intergranular crack path is highly favoured compared to the transgranular crack path (Fig.5), so that the intergranular fracture mode dissipates less energy than the transgranular mode [36].

The impact energy for dissimilar welds was slightly higher, but of the same order, than the MMC-MMC welds, because the fracture path was mainly located in the composite side, as discussed in the fracture surface section, contributing to the low values of the total absorbed energy. The percentage of W_m is indeed in the range of 60÷75 % of the total energy, both for the MMC samples, welded and unwelded, and for the dissimilar Al-MMC joints.

A comparison of representative load-displacement curves, from the instrumented impact tests, for the MMC-MMC, Al-Al and Al-MMC LFW samples is reported in Fig.8. The curve of the Al-Al LFW sample shows the lowest ratio between the nucleation and the propagation energies, and the highest value of the peak force F_m. Its low rate of load reduction after the peak force, indicates a controlled crack propagation. Moreover, as can be seen in Table 3, for both the AA2024 Al alloy and AA2124/25vol.%SiC_p composite, the average maximum force decreases of about 9 % in the LFW samples with respect to the base metal. The effect of LFW and, consequently, of microstructural modifications on the dynamic stresses, seem to be not negligible.

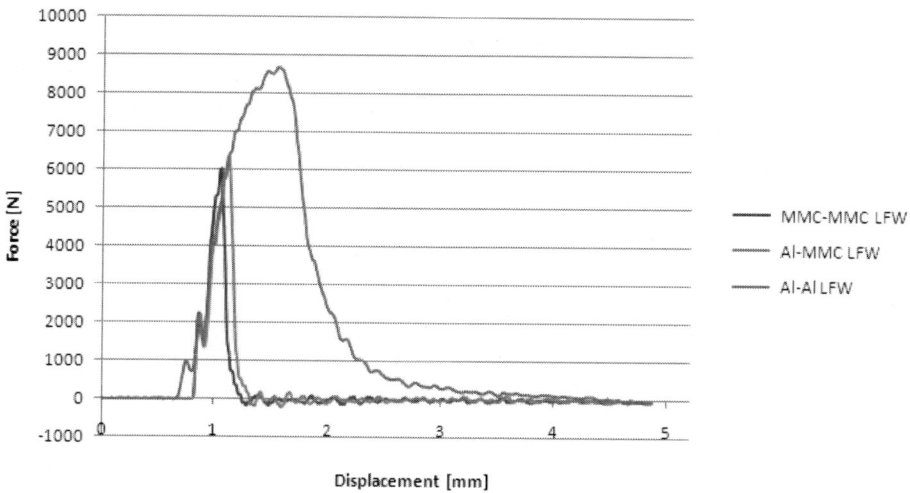

Figure 8. Representative load-displacement curves from the instrumented impact tests for: the AA2124 MMC (MMC-MMC LFW), AA2024 (Al-Al LFW) and dissimilar joints (Al-MMC LFW).

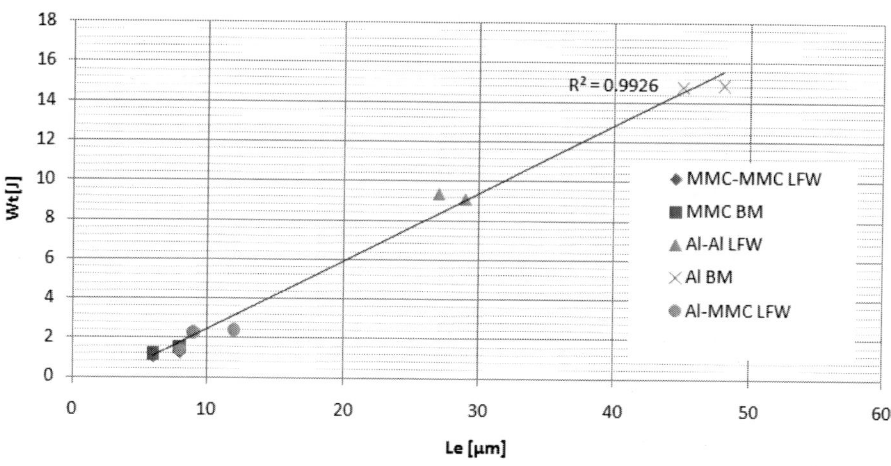

Figure 9. Total impact energy versus lateral expansion for the MMC welded and base material, AA2024 base and welded alloy, Al-MMC dissimilar joints.

The lateral expansion measurements were found to be in excellent agreement with the measured impact energies. In particular, a linear correlation was found with the total impact energy, as shown by the plot in Fig.9, suggesting how lateral expansion measurements could also be used as a criteria to assess the impact behaviour of Al alloys and Al-based composites.

The low magnification micrographs of the fracture surfaces of the LFW AA2024 Al alloy (Fig.10-a and -c) show a line pattern, parallel to the crack growth direction, which reflects the fibrous grain structure. In the base AA2024 specimens (Fig.10-b and -d) no such line pattern is visible, because the notch and thus the crack growth direction, are normal to the fibrous grains (Fig.10-b) and the crack front profile is rougher in respect to that of the LFW samples (Fig.10-d).

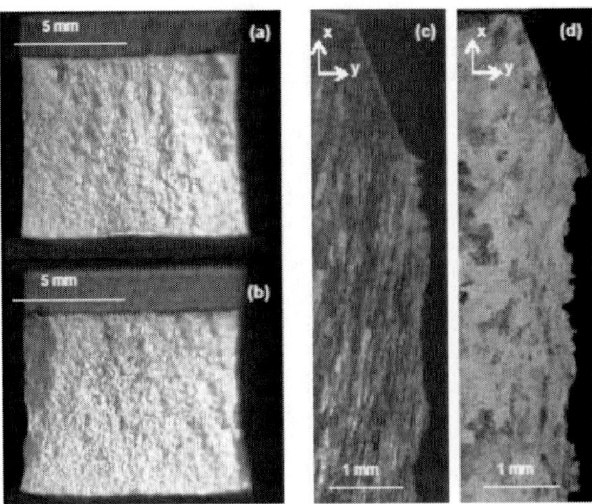

Figure 10. (a, b) Stereomicroscopy images of fracture surfaces and (c, d) optical images of fracture sections of (a, c) the welded Al-Al LFW and of (b, d) the base AA2024.

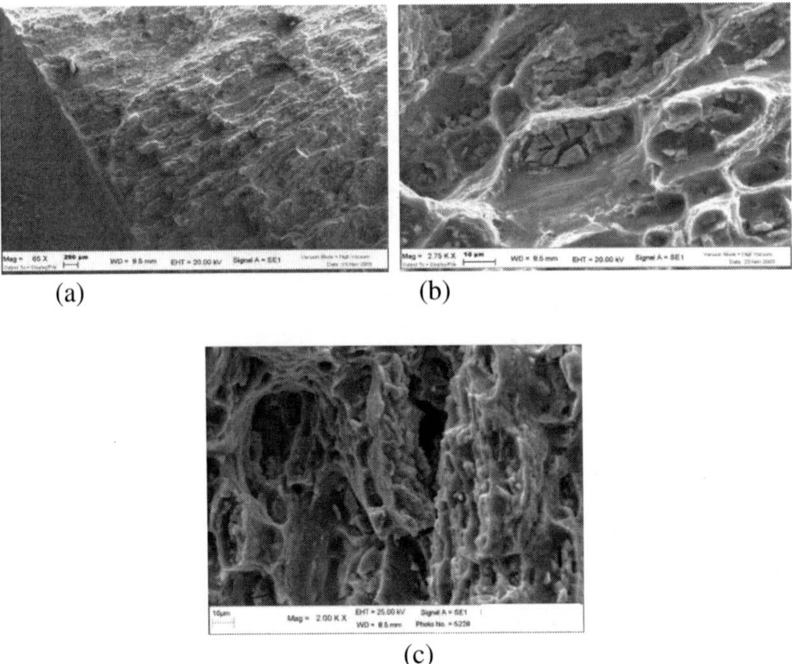

(a) (b)

(c)

Figure 11. SEM images of the fracture surfaces of (a, b) the Al-Al LFW and (c) base AA2024 alloy.

SEM analyses of the fracture surfaces of the AA2024 LFW samples confirmed that the fracture path follows the fibrosity of the material (Fig.11-a); however, at higher magnification, welded (Fig.11-b) and base (Fig.11-c) 2024 Al alloys displayed similar morphologies, typical of a ductile failure. Fracture surfaces were characterised by larger

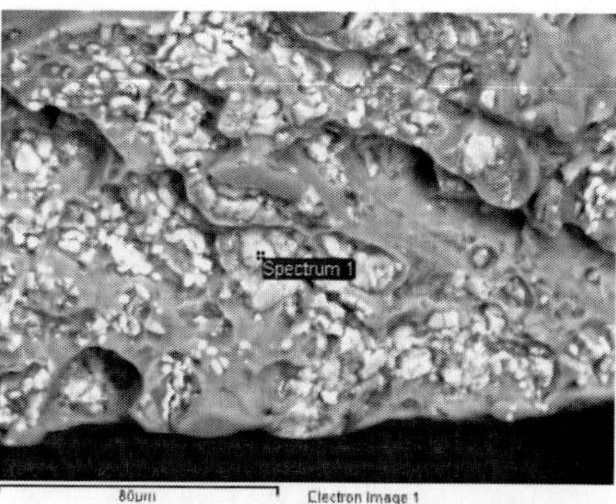

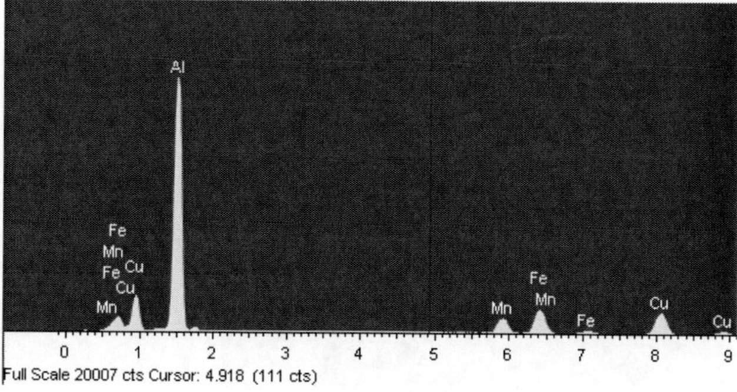

Figure 12. EDS analysis on the compounds found inside the smaller dimples in the AA2024 alloy.

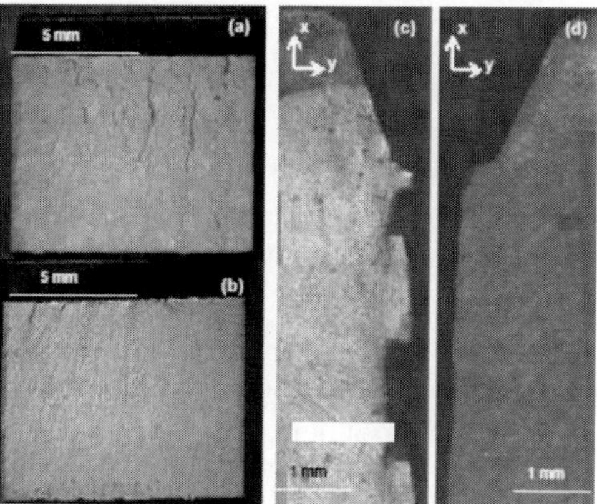

Figure 13. (a,b) Stereomicroscopy images of fracture surfaces and (c, d) optical images of fracture sections of (a, c) the welded MMC and of (b, d) the base MMC.

dimples, with a faceted appearance, surrounded by smaller dimples, although some intergranular secondary cracks were also visible. Moreover, a large amounts of intermetallic compounds, some of which were broken, were visible at many locations, especially inside the smaller dimples. The EDS analysis suggested that these particles were a complex compound of Fe, Mn, Si and Al (Fig.12) and, to a lesser degree, CuAl2 particles [37].

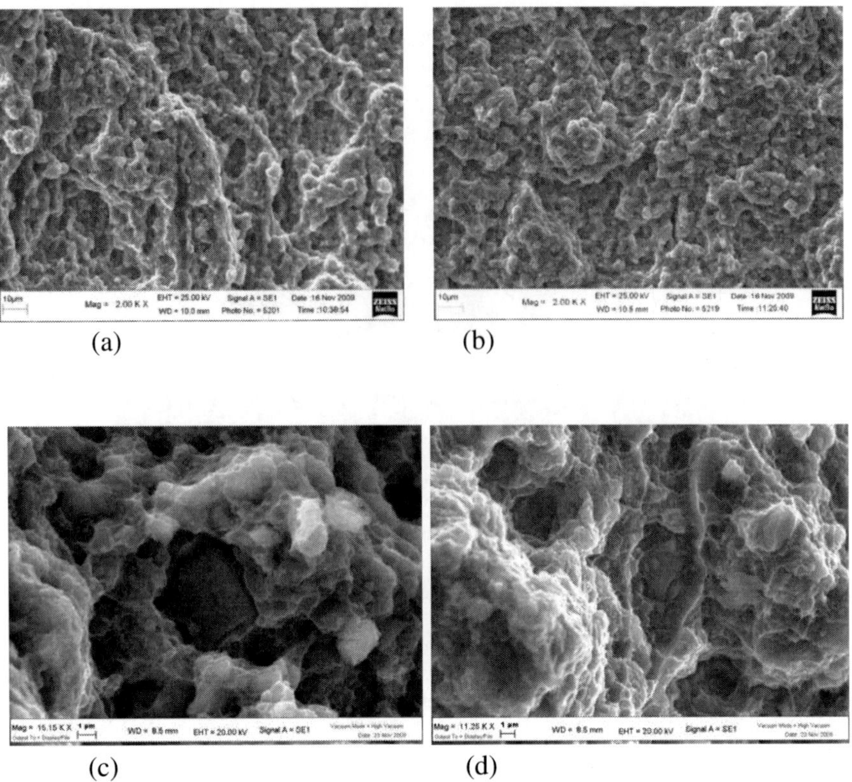

Figure 14. SEM micrographs of the fracture surfaces of (a, c) the MMC-MMC LFW and (b, d) base MMC.

The low magnification micrographs of the fracture surfaces of the MMC-MMC LFW do not clearly show a fracture line pattern, even if it propagates on different planes (Fig.13-a and -c) following the weakest path induced by the LFW process. In the base material (AA2124/SiC_p composite), the fracture appears to be flatter (Fig.13-b and -d) than in the LFW joint.

Nevertheless, SEM analyses at high magnification emphasized similar morphology in the welded (Fig.14-a and -c) and base composite (Fig.14-b and -d), showing the presence of secondary cracks and a bimodal distribution of voids: the larger ones associated with decohesion of the reinforcement particles, and the smaller ones associated with the ductile failure of the Al alloy matrix. Due to the small size of the reinforcing particles, only a small amount of cracked particles was detected and no brittle phases were observed at the interfaces due to undesired chemical reactions (Fig.14).

The low magnification micrographs of the fracture surfaces of the dissimilar AA2024/MMC LFW joints (Fig.15-a and -b) showed two different morphologies: the first one was characterized by a line pattern reproducing the fibrous grain structure induced by the LFW; the second one appeared completely flat, with no visible line pattern.

High magnification SEM analyses (Fig.15-c and -d) highlighted: (i) the classic morphology of the fracture surfaces of MMCs, clearly associated with the crack propagation

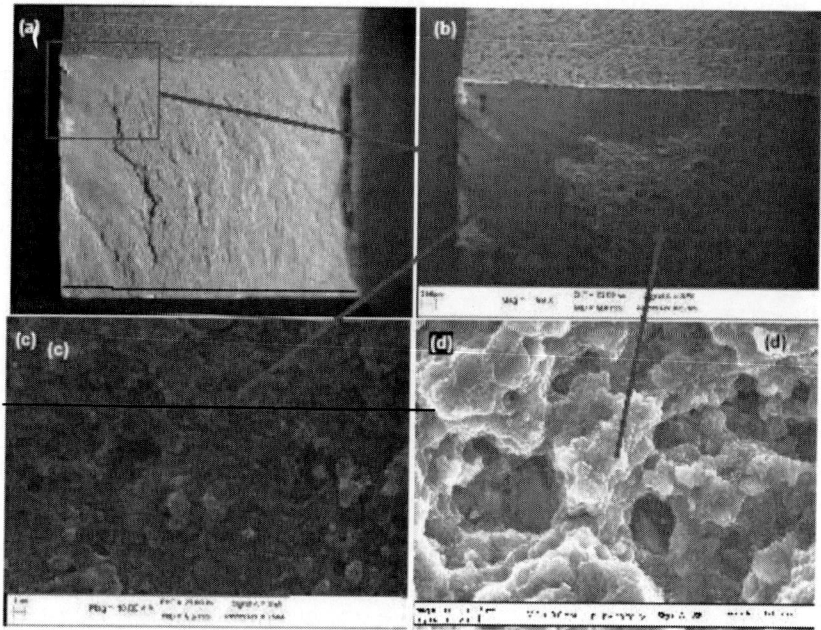

Figure 15. Representative low (a-b) and high (c-d) magnification SEM micrographs of the fracture surface of the dissimilar Al-MMC LFW joints. Ultra-fine dimples (d<1 μm) in correspondence of (c) the weld centre of the AA2024 side, (d) morphology of MMCs fracture surfaces.

in the MMC side of the joint; (ii) the presence of ultra-fine dimples (d<1μm), which are probably related to the fine microstructure of the Al alloy, in the weld centre.

These results, in accordance with the impact data that show a similar behaviour of MMC-MMC and Al-MMC LFW joints, confirm that in dissimilar joints a crack develops preferentially in the TMAZ of the MMC or along the boundary between the Al alloy and the MMC.

Conclusion

Instrumented impact tests were carried out on similar and dissimilar joints, obtained by Linear Friction Welding, between a AA2024 Al alloy and a SiC particle reinforced Al based composite (MMC). Three characteristic zones were identified in all the analyzed LFW specimens: a central zone, with an ultrafine microstructure (and uniform particle distribution in the case of MMC); a thermo-mechanically affected zone (TMAZ) with severe plastic deformation; a heat affected zone (HAZ), without visible microstructural modification of the material.

An interesting result was that the average total impact energy of the LFW composite was substantially similar to that of the base MMC, suggesting that, contrary to fusion welding processes, this solid state welding process does not affect the impact properties. This is clearly due to the fact the LFW MMC joints showed a microstructure substantially comparable with that of the base material, in terms of reinforcement particle size and distribution.

The LFW, on the contrary, had a higher influence on the impact behaviour of the unreinforced AA2024 Al alloy, causing a decrease of about 37 % in the total absorbed energy. This could be due to the elongated grain texture induced by LFW, which being parallel to the crack nucleation and propagation direction, reduced both the nucleation and propagation energies. The impact energy for dissimilar welds was slightly higher, but of the same order, than the MMC-MMC welds, because the fracture path was mainly located in the composite side.

REFERENCES

[1] Clyne TW, Withers PJ. *An introduction to metal matrix composites.* Cambridge University Press; 1993.

[2] Taya M, Arsenault RJ. *Metal matrix composites – thermomechanical behaviour.* Oxford: Pergamon Press; 1989.

[3] Lloyd DJ. *Particle reinforced aluminium and magnesium metal matrix composites.* Int Mater Rev 1994;39(1):1–45.

[4] Torralba JM, Da Costa CE, Velasco F. *P/M aluminum matrix composites: an overview.* J Mater Process Technol 2003(1–2):203–6.

[5] Subramanian KN, Bieler TR, Lucas JP. *Mechanical shaping of metal matrix composites.* Key Eng Mater 1995;104–107:175–86.

[6] Davies CHJ. *Critical issues in the extrusion of particle reinforced metal matrix composites.* Key Eng Mater 1995;104–107:447–56.

[7] Urena A, Escalera MD, Gil L. *Influence of interface reactions on fracture mechanisms in TIG arc-welded aluminium matrix composites.* Compos Sci Technol 2000;60:613–22.

[8] Ellis MBD. *Joining of aluminium based metal matrix composites.* Int Mater Rev 1996;41(2):41–58.

[9] Zhang XP, Quan GF, Wei W. *Preliminary investigation on joining performance of SiCp-reinforced aluminium metal matrix composite (Al/SiCp–MMC) by vacuum brazing*. Comp Part A 1999; 30(6):823-827.

[10] Wang HM, Chen YL, Yu LG. *'In-situ' weld-alloying/laser beam welding of SiCp/6061Al MMC.* Mater Sci Engineering A 2000; 293(1-2):1-6.

[11] Ureña A, Escalera MD, Gil L. *Influence of interface reactions on fracture mechanisms in TIG arc-welded aluminium matrix composites.* Comp Sci Tech 2000; 60(4):613-622.

[12] Lean PP, Gil L, Ureña A. *Dissimilar welds between unreinforced AA6082 and AA6092/SiC/25p composite by pulsed-MIG arc welding using unreinforced filler alloys (Al–5Mg and Al–5Si).* J. Mater Process Technol 2003; 143-144:846-850.

[13] Chen CM, Kovacevic R. *Joining of Al 6061 alloy to AISI 1018 steel by combined effects of fusion and solid state welding*, International Journal of Machine Tools and Manufacture. 2004; 44(11):1205-1214.

[14] Zhang XP, Ye L, Mai YL, Quan GF, Wei W. *Investigation on diffusion bonding characteristics of SiC particulate reinforced aluminium metal matrix composites (Al/SiCp-MMC).* Comp Part A. 1999; 30(12):1415-1421.

[15] Mishra RS, Ma ZY. *Friction stir welding and processing,* Mater Sci Eng. 2005; 50(1-2):1-78.

[16] Gaafer AM, Mahmoud TS, Mansour EH. *Microstructural and mechanical characteristics of AA7020-O Al plates joined by friction stir welding, Mater Sci Eng A*, In Press, Corrected Proof, Available online 19 August 2010.

[17] Khodir SA, Shibayanagi T. *Friction stir welding of dissimilar AA2024 and AA7075 aluminum alloys,* Mater Sci Eng: B 2008; 148(1-3):82-87.

[18] Fernandez GJ, Murr LE. *Characterization of tool wear and weld optimization in the friction-stir welding of cast aluminium 359 + 20% SiC metal–matrix composite.* Mater Charact 2004;52:65–75.

[19] Wert JA. *Microstructures of friction stir weld joints between an aluminium base metal matrix composite and a monolithic aluminium alloy.* Scripta Mater 2003;49:607.

[20] Boromei I, Ceschini L, Morri A, Garagnani G. *Friction stir welding of aluminium based composites reinforced with AL2O3 particles: effects on microstructure and Charpy impact energy.* Metall Sci Technol 2006;24:12–21.

[21] Ceschini L, Boromei I, Minak G, Morri A, Tarterini F. *Microstructure, tensile and fatigue properties of AA6061/20 vol.%Al2O3p friction stir welded joints.* Composites Part A 2007;38:1200–10.

[22] Ceschini L, Boromei I, Minak G, Morri A, Tarterini F. *Effect of friction stir welding on microstructure tensile and fatigue properties of the AA7005/10 vol.% Al2O3p composite.* Compos Sci Technol 2007;67:605–15.

[23] Minak G, Ceschini L, Boromei I, Ponte L. *Fatigue properties of friction stir welded particulate reinforced aluminium matrix composites.* Intern J of Fatigue 2010; 32(1):218-226.

[24] Liu HJ, Feng JC, Fujii H, Nogi K. *Wear characteristics of a WC–Co tool in friction stir welding of AC4A+30 vol%SiCp composite*. Int J of Machine Tools and Manuf. 2005; 45(14):1635-1639.

[25] Vairis A, Frost M. *High frequency linear friction welding of a titanium alloy.* Wear 1998; 217:117–31.

[26] Vairis A, Frost M. *On the extrusion stage of linear friction welding of Ti 6Al 4V.* Mater Sci Eng 1999; A271:477–84.

[27] Vairis A, Frost M. *Modelling the linear friction welding of titanium blocks.* Mater Sci Eng A 2000; 292:8–17.

[28] Li WY, Ma TJ, Yang SQ, Xu QZ, Zhang Y, Li JL. *Effect of friction time on flash shape and axial shortening of linear friction welded 45 steel.* Mater Lett 2008;62(2):293–6.

[29] Karadge M, Preuss M, Withers PJ, Bray S. *Importance of crystal orientation in linear friction joining of single crystal to polycrystalline nickel-based superalloys.* Mater Sci Eng A 2008;491(1–2):446–53.

[30] Jun TS, Rotundo F, Song X, Ceschini L, Korsunsky AM. *Residual strains in AA2024/AlSiCp composite linear friction welds.* Materials and Design, Volume 31, Supplement 1, June 2010, Pages S117-S120

[31] Rotundo F, Ceschini L, Morri A, Jun TS, Korsunsky AM, *Mechanical and microstructural characterization of 2124Al/25 vol.%SiCp joints obtained by linear friction welding (LFW),* Composites: Part A 2010;41:1028–1037.

[32] ASTM E3-01. *Standard practice for preparation of metallographic specimens.* ASM International; 2007.

[33] ASTM E23-07a *Standard Test Methods for Notched Bar Impact Testing of Metallic Materials*, ASM International; 2007.

[34] UNI EN ISO 14556:2003 *Steel - Charpy V-notch pendulum impact test* – Instrumented test method; 2003.

[35] Ozden S, Ekici R, Nair N, *Investigation of impact behaviour of aluminium based SiC particle reinforced metal–matrix composites*, Composites: Part A 2007;38:484–494.

[36] Chena Y, Pedersen KO, Clausen AH, Hopperstad OS. *An experimental study on the dynamic fracture of extruded AA6xxx and AA7xxx aluminium alloys*. Mater Sci Engineering A 2009;523:253–262.

[37] Kaçar H, Atik E, Meriç C.*The effect of precipitation-hardening conditions on wear behaviours at 2024 aluminium wrought alloy*. J of Mater Process Technol 2003;142(3):762-766.

In: Metal Matrix Composites
Editor: J. Paulo Davim

ISBN: 978-1-61209-771-8

Chapter 7

DRILLING OF METAL MATRIX COMPOSITES: A BRIEF OVERVIEW

S. Basavarajappa*[*1], M. Prasannakumar[2] and J. Paulo Davim[3]

[1]Dept. of Studies in Mechanical Engg.,
University B.D.T. College of Engineering,
Davanagere University, Karnataka, India
[2]Dept. of Mechanical Engineering,
Bapuji Institute of Engineering and Technology, Karnataka, India
[3]Dept. of Mechanical Engineering, University of Aveiro,
Campus Santiago, Aveiro, Portugal

ABSTRACT

The use of metal matrix composite materials has been increased considerably over the last decade and, as a consequence, more number of researchers are working in this area. The objective of this chapter is to present a brief literature review on drilling of theses materials. Aspects such as tool materials and geometry, machining parameters and their influence on the thrust force, surface roughness, Burr formation are reviewed based on SiCp, Graphite and SiCp-Graphite reinforced Aluminium matrix composites. Additionally, the subsurface alterations and statistical analysis after drilling on these materials were discussed. This allows the readers to a better understanding of the material behaviour and drillability characteristics of this category of materials.

[*]Corresponding author, E-mail: basavarajappas@yahoo.com

1. Metal Matrix Composites

The word '*composite*' in composite materials signify that two or more materials are combined on a macroscopic scale to form a useful material. Different materials can be combined on a macroscopic scale resulting to a macroscopically homogeneous material. The advantages of the composites are that they usually exhibit the best qualities of their constituents. Generally, composite materials will consist of two separate components, the matrix and the reinforcement. The matrix is the component that holds the reinforcement together to form the bulk of the material.

The reinforcement is the material that has been reinforced in the matrix to lend its advantage (usually strength) to the composite. The reinforcements can be of any material such as carbon fiber, glass bead, sand or ceramics et al. The properties that can be improved by forming a composite material includes, high specific strength and modulus, wear resistance, corrosion resistance, attractiveness, fatigue life, temperature-dependent behaviour, thermal insulation [1, 2]. The composites were classified broadly based on matrix materials are (i) Polymer matrix composites (ii) Metal matrix composites (iii) Ceramic matrix composites.

Conventional monolithic materials have limitations in terms of achievable combinations of strength, stiffness, coefficient of expansion and density. Metal Matrix Composites (MMCs) have emerged as an important class of advanced materials giving engineers the opportunity to tailor the material properties according to their needs. A Metal matrix composite is an engineered combination of two or more materials (one of which is a metal) in which tailored properties are achieved by systematic combination of different constituents. Development of these materials is a subject of great interest as they offer attractive combination of physical and mechanical properties, which cannot be obtained in monolithic alloys. Essentially, these materials differ from the conventional engineering materials from the point of homogeneity [3].

The major advantages of MMCs compared to unreinforced materials are as follows:

- Higher strength-to-density ratios
- Higher stiffness-to-density ratios
- Better fatigue resistance
- Better elevated temperature properties
- Lower coefficients of thermal expansion
- Improved abrasion and wear resistance
- Improved damping capabilities
- Improved corrosion resistance

2. Machining of Metal Matrix Composites

Despite the superior mechanical and thermal properties of Discontinuously Reinforced Particulate Metal Matrix Composites (DRMMCs), their poor machinability has been the main deterrent to their substitution for conventional metal parts. The hard abrasive reinforcement phase in the soft matrix causes rapid tool wear during the machining and consequently, high machining costs. It is clear that the morphology, distribution and volume fraction of the reinforcement phase as well as matrix properties are all factors that affect the overall cutting process, but as yet only a relatively few works relating to the optimization of the machining process have been published.

Despite the controversy in explaining the mechanism behind the tool wear at different feed rates, all reviewed literature recommend using feed rates and depths of cut that are as aggressive as possible during the roughing operations. Several researchers have indicated that polycrystalline diamond (PCD) tools are the only tool material that is capable of providing a useful tool life during the machining of SiC/Al PMMCs [4]. PCD is harder than Al_2O_3 as well as SiC and does not have a tendency to react chemicals with the work piece material. Furthermore, PCD tools with a grain size of 25 μm will withstand abrasion wear by micro-cutting better than tools with a grain size of 10 μm. Non-conventional machining processes, such as Electro Discharge Machining (EDM), laser cutting and Abrasive Water Jet (AWJ), have also been reviewed and found to be highly suitable for rough cut application.

3. Drilling

A major obstacle in the development of MMCs as a replacement for monolithic alloys is its processing characteristics. Drilling cannot be avoided and is required for finishing the component to a final size and for the final assembly. The applications of MMCs are limited owing to their poor drillability, which is a result of the highly abrasive nature of the reinforcing particles. These abrasive particles cause excessive tool wear during cutting of ceramic reinforced metal matrix composites with tungsten carbide tools and even with diamond tools. Attempts have been made to eliminate drilling operations by using fabrication techniques such as near net shape forming and modified casting techniques. These techniques however, having their own limitations and therefore drilling is still an integral part of the component manufacture for the final assembly.

By controlling the drilling parameters, the intended surface finish, reduced tool wear, reduced burr both in the entry and exit and reduced cutting forces can be obtained. From the available literature on drilling of MMCs, it is obvious that the morphology, distribution and volume fraction of the reinforcement phase as well the matrix properties are the factors controlling the overall drilling properties. This chapter describes the effect of drilling parameters, drilling tools, surface and subsurface damage of drilled parts of Al/SiCp, Al/Gr and Al/SiCp-Gr composites.

3.1 Drilling of SiCp Reinforced Composites

During drilling of MMCs, the cutting tool undergoes severe abrasive wear due to the presence of hard reinforcements and is considered as highly unproductive [5].

Many researchers have studied the drilling characteristics of ceramic reinforced composites [6, 7]. Ramulu et al [8] conducted drilling experiments on Al_2O_3 reinforced aluminium based metal matrix composites using different drill materials. They found that the abrasive particles cause excessive flank wear and HSS tools are not suitable for machining these materials. Coelho et al [5] in their study compared the reaming of Al/SiC MMCs drilled with HSS, WC, TiN coated WC and PCD drills. They recommended WC and TiN coated WC can be used for batch production. PCD drills are the only realistic tooling option in large batch and mass production. The rating is based upon the number of holes drilled before a tool wear criterion is achieved.

Davim and Baptista [9] presented an experimental study on drilling of aluminum matrix composites using PCD drills. The surface roughness of the drilled holes was evaluated and obtained very good results. Mubaraki et al [10] examined the wear behaviour of HSS, WC and PCD tools in drilling of Al_2O_3 reinforced Al alloy particulate metal matrix composites, and established a correlation between the flank wear and the cutting forces.

Monaghan et al. [11] Reported when drilling SiCp reinforced aluminium matrix composites. They attributed the improved surface finish to the burnishing or honing effect produced by the action of small SiC particles trapped between the flank face of tool and the workpiece surface. Most of the researchers recommended the study on the effect of the cutting parameters on the sub-surface deformation produced due to high stress and temperatures generated during cutting as well as to the burr formation mechanism.

Burr formation is an important aspect in drilling of MMCs. The burr formation in machining can be divided into three stages namely, initiation, development, and formation of burr. It is also obvious from the literature that workpiece material, drill geometry/material, amount of built up edge and feed rate are the predominant factors influencing the variation in burr size [12]. Gillespie and Blotter [13] differentiated the burr based on its formation mechanism as Poisson burr, Roll over burr, Tear burr and Cut off burr. The burr formed in drilling operation falls under the roll over burr type. The roll over burr is essentially a chip that is pushed out of the cutter's path rather than sheared. The burr formation process starts at a transition point where natural chip formation stops and plastic deformation below the plane of machined surface in the axial direction begins.

Iwata et al [14] have reported the plastic deformation is characterised by a negative shear angle and negative shear plane. This negative shear plane is considered to extend towards the workpiece edge and the negative shear angle is characterized by cutting direction. As the plastic deformation increases in the negative shear plane, catastrophic deformation occurs at the edge of the workpiece. As the drill advances, the catastrophic deformation expands and connects with primary shear zone. With further advancement due to larger deformation, the cutting force decreases and the negative shear plane rotates with the pivot point as plastic hinge [15,16].

The initiation and development stages in burr formation are not significantly influenced by the material properties. The crack initiation at the tool tip and propagation leading to burr. In a ductile material, the crack is along the cutting line due to the larger critical fracture strain of ductile material that is not available in negative shear plane. In a brittle material, the crack propagates towards the pivot point and it leads to break out. The drill geometry employed has a noticeable influence in the burr formation. The elements of drill shape, point angle, helix angle and length of chisel edge also influence burr height. The volume of the material that undergoes bending depends on the drill shape employed. When a multifaceted drill is used, the burr initiation is delayed and shearing action is prolonged by the change in the cutting action and the height of burr is reduced.

Barnes and Pashby [17] have studied and compared the burr formation in Al 2618/18SiCp MMCs during dry, conventional and through coolant drilling. They are of the opinion that conventional coolant supply does not reduce the burr height, but through coolant supply reduces the burr height by 25 percent.

3.2 Drilling of Graphitic Composites

Most of the current literature presents experimental results on drilling of ceramic reinforced MMCs. However, very marginal information is available on drilling of graphitic reinforced composites. Sharma et al [18] have explained that the tool life of HSS drill is increased while machining the aluminium graphite composite compared to the base alloy. There is a reduction in energy required for drilling the graphitic composite compared to the base alloy since graphite being a solid lubricant reduces the friction at the tool-work interface.

In their further work Songmene et al [19] have investigated the drilling of these composites as compared to un-reinforced aluminium. The microstructure of chip, the cutting forces, the shear angles and the friction at tool-chip interface are used to compare the machinability of these composites. It was found that, during drilling of this new family of composites, the feed rate, and the nature of reinforcing particles govern the cutting forces. Mathematical models established by previous researchers for predicting the cutting forces when drilling metals were found valid for these composites. The reinforcing particles within the composite help for chip segmentation, making the composite more brittle and easy to shear during the cutting process.

3.3 Drilling of Hybrid Composites

Most of the literature presents the experimental results when drilling ceramic reinforced or graphitic reinforced composites. Limited literature is available on drilling of Ceramic-graphite reinforced hybrid metal matrix composites. Songmene and Balazinzki [20] worked on drilling and milling of Al/SiCp, Al/SiCp–Gr and Al/Al2O3–Gr composite and they are under the opinion that the incorporation of graphite particle into aluminium MMCs and the variation of hard particle content improve the machinibility of the composite. The torque when drilling graphitic composites was comparable to that of Al 380 aluminium alloy.

Basavarajappa et al [21] reported while drilling Al 2219/SiCp andAl219/SiCp-Gr hybrid metal matrix composites with various cutting speeds and federates. The Fig.1 and Fig.2 shows the Variation Thrust force with feed rate and cutting speed for carbide and coated carbide tools for both Al/SiCp and Al/SiCp-Gr composites.

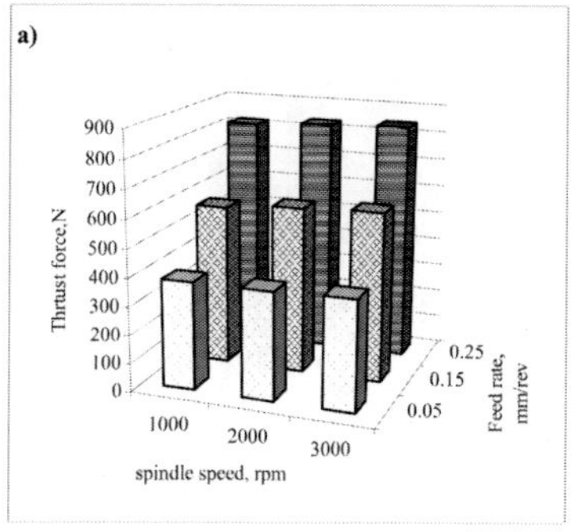

Figure 1. (continued)

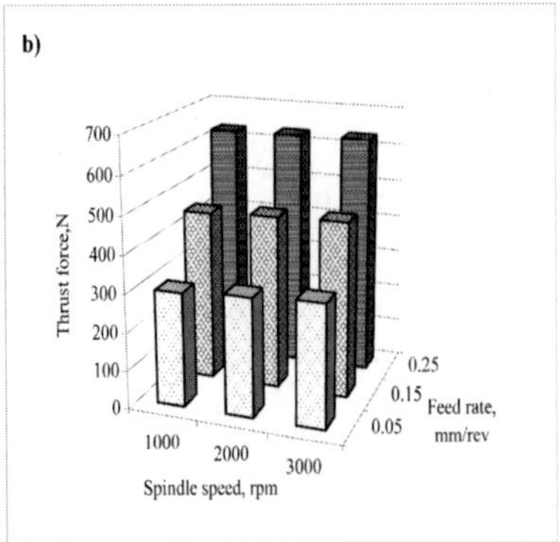

Figure 1. Thrust force induced by solid carbide multifaceted drills for (a) Al2219/15SiCp composite and (b) Al2219/15SiCp-3Gr composite.

It shows that the Al/SiCp-Gr composites shows the less thrust force compared to SiCp reinforced aluminium composites. The lesser thrust force attributed to the solid lubricating property of the graphitic composites.

The graphite particles reduce the interfacial friction between the tool and the workpiece and lower the shear flow stress. The addition of graphite
reduces the thrust force by almost 30% when drilling with coated carbide tools and 20% when drilling with carbide cutting tools. The coated carbide tool performed better than the carbide tool in all the cutting conditions.

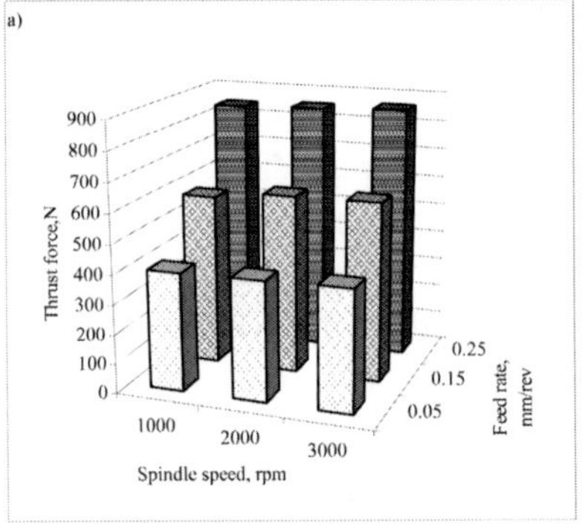

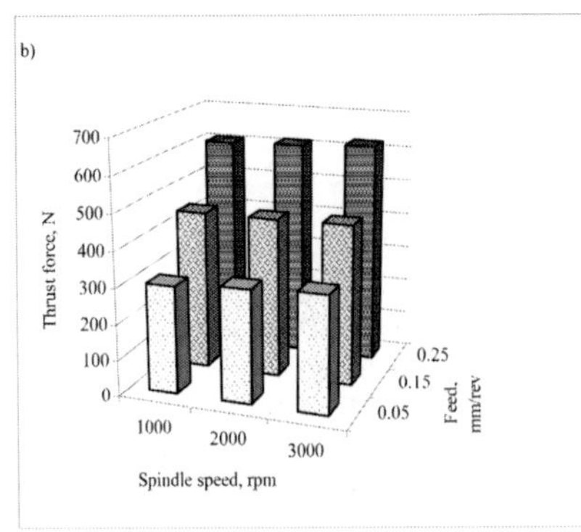

Figure 2. Thrust force induced by coated carbide for a) Al2219/15SiCp composite and b) Al2219/15SiCp-3Gr composite.

Fig. 3 and Fig.4 shows the variation of surface roughness with different cutting conditions with carbide and coated carbide tools for both SiCp and SiCp-Gr reinforced aluminium composites. They reported that the surface roughness values of the graphitic composites are relatively more when compared to SiCp reinforced composites for all cutting conditions and for both the type of drill bits. The surface roughness value of the drilled surface decreases with an increase in speed for all cutting conditions and for both the tools and workpiece materials. This is due to the burnishing or honing effect produced by the rubbing of small SiC particles trapped between the flank face of the tool and the workpiece surface. For all cutting conditions, Al2219/15SiCp composite has less surface roughness than Al2219/15SiCp-3Gr composite. The higher surface roughness values for Al2219/15SiCp-3Gr composite can be attributed to their easy pullout from the surface. Hence, deep valleys are formed at the surface of the composite resulting in poor surface finish [20]. The surface finish is good for coated carbide drill when compared to multifaceted carbide drill. Hence, usage of coated carbide tool is advantageous in applications that require good surface finish.

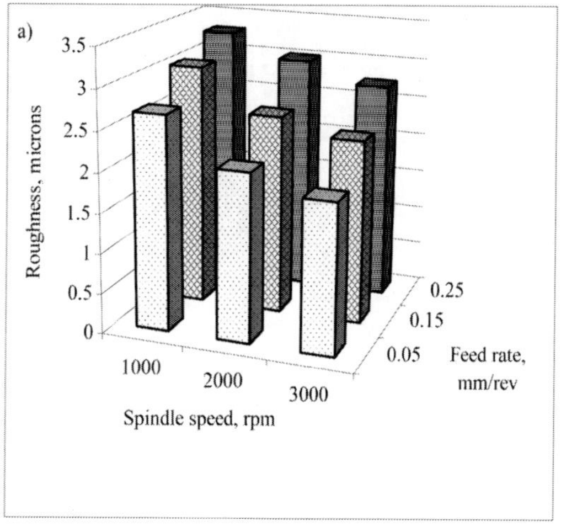

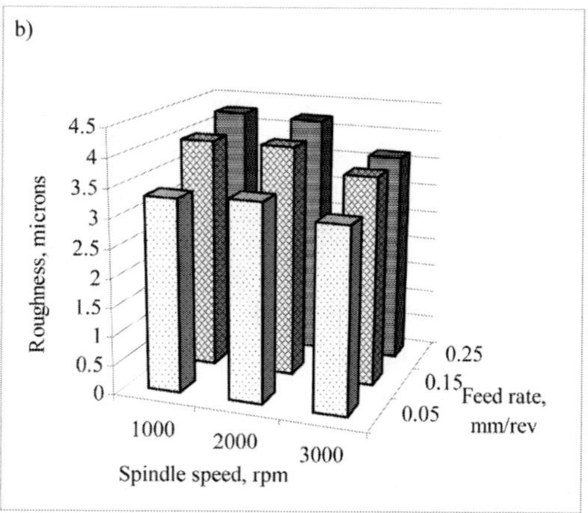

Figure 3. The surface roughness values of the composites drilled by solid carbide multifaceted drills at various spindle speeds and feed rates a) Al2219/15SiCp composite b) Al2219/15SiCp-3Gr composite.

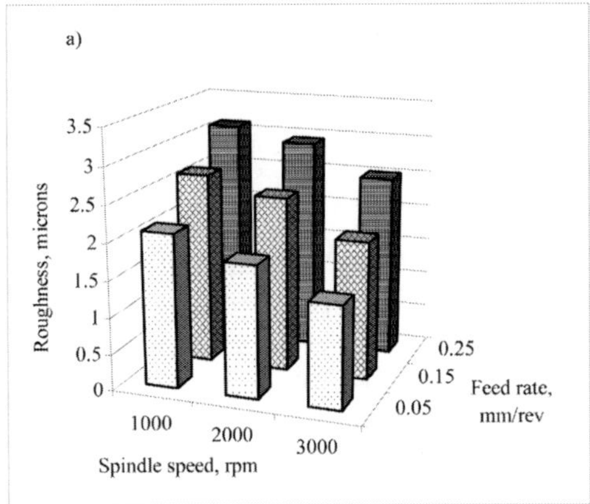

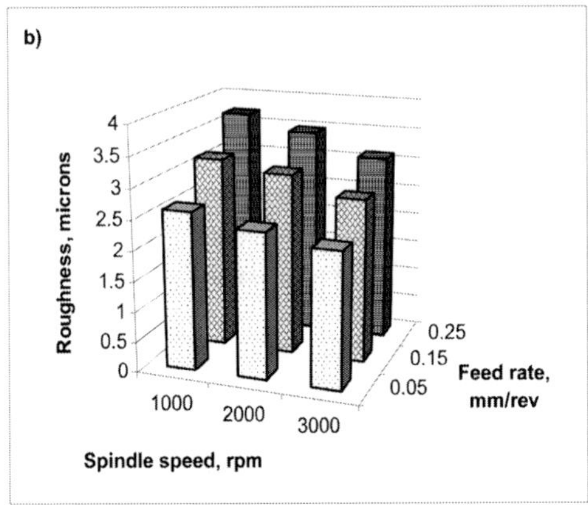

Figure 4. The surface roughness values of the composites drilled by coated carbide drills at various spindle speeds and feed rates a) Al2219/15SiCp composite. b) Al2219/15SiCp-3Gr composite.

4. Subsurface Deformation in Drilling of Composites

Drilling is carried out in the final processing stage before assembly and the drilled holes act as stress concentration locations. Hence, increased attention must be paid to the possibilities of service failures caused by creep and fatigue. The effect of machining on the type of surface defects and their distribution has a major impact on the performance of the machined component[22].

Most failures in service originate at or near the surface, therefore maintenance of surface in turn leads to development and maintenance of structural integrity [23, 24,]. Novovic et al [25] have studied the effect of machining and the resulting workpiece surface topography/integrity on fatigue performance, for a variety of workpiece materials. They found that residual stress is one of the important factors that influences fatigue life of the component and this effect greatly depends on material property and strain hardening.

In metal removal process, the interaction of the tool and workpiece may create a Disturbed Material Zone (DMZ) and causes various geometrical and metallurgical defects in the surface region like fine cracks and high residual stresses [22,23]. Tosun and Muratoglu [26] analyzed the drilled surface and subsurface of Al 2124/17 SiCp MMC and found that the effect of point angles on sub-surface damage caused by drilling operation changed with the type of drills. They also found that the increase in hardness of the tool material decreases the surface roughness of drilled surface.

El-Gallab and Sklad [27] studied the effect of cutting parameters on the surface quality and the extent of sub-surface damage in machining of Al/20SiCp MMCs. They found that surface roughness increases with the increase in cutting speed and subsurface damage extends up to 100 microns beneath the machined surface.

Basavarajappa el al [28] reported that the addition Graphite in SiCp reinforced Aluminium matrix composites. In the Fig. 5 the subsurface deformation extends up to a maximum of 120 μm below the machined surface for Al2219/15SiCp-3Gr composite when compared to 150 μm in Al2219/15SiCp composite. Multifacet carbide drill performed better than conventional coated carbide drill in terms of subsurface deformation.

5. Statistical Analysis in Drilling of Composites

Taguchi design of experiments and analysis of variance was successfully used in machining process to access the influence of the various factors and its interaction affecting the machinability of the materials.

Davim and Antonio [29] have conducted drilling tests with the intention of developing optimal drilling conditions using genetic algorithm approach. The surface finish is found to be affected by the feed rate and not by the cutting speed. Davim [30] studied the drilling of metal matrix composites based on the Taguchi technique to find the influence of cutting parameters on tool wear, torque and surface finish and the interactions of the above factors. He analyzed the data by analysis of variance and found the percentage of influence of each factor on responses. Ramulu et al [31] conducted experiments by using PCD drills to drill Al_2O_3 particulate reinforced aluminium based metal matrix composites. The analysis of variance (ANOVA), Response surface methodology were used to analyze experimental data and developed regression models. They concluded that drilling forces and average surface roughness values are greatly influenced by the feed rate than the speed.

Conclusion

Optimization of machining parameters (speed, feed,) for a better surface finish, tool life, minimum cutting forces, Minimum burr formation and Minimum subsurface alterations greatly depends on the matrix material, reinforcement type, volume fraction, and size of the reinforcement. The ceramic particulates (SiCp, Al_2O_3, etc.) incorporated in the metal matrix composites are comparatively harder than the commonly used cutting tools. The cutting tools recommended to machine these materials in the order of preference are PCD, CBN, TiC, Si_3N_4,

Al_2O_3 and WC. The cutting forces are minimum with the addition of Graphite and the surface roughness is more. Tool wear increases with an increase in cutting speed, particulate size, and volume fraction. PCD tools show a better tool life at all cutting conditions compared to CVD and coated carbide tools.

REFERENCES

[1] Ibrahim A., Mohammed F.A. and Lavernia E.J. *Metal matrix composites- A review,* Journal of Material Science 1991, 26, 1137-1157.

[2] Hooker J.A. and Doorbar P.J. *Metal matrix composites for aero engines,* Materials Science and Technology 2000, 16, 725-731.

[3] Taya M. and Arsenault R.J. *Metal matrix composites-* Thermo mechanical behavior, Pergamon press 1989, New York.

[4] Davim J. Paulo. *Diamond tool performance in machining metal-matrix composites,* Journal of Material Processing Technology 2002, 128, 100-105.

[5] Coelho R.T., Yamada S., Aspinwall D.K. and Wise L.H. *The Application of polycrystalline diamond (PCD) tool materials when drilling and reaming aluminum based alloys including MMC,* International Journal of Machine Tools and Manufacturing 1995, 35, 761-774.

[6] Tosun Gul and Muratoglu Mehtap. *The drilling of Al/SiCp metal-matrix composites.* Part I: microstructure, Composite Science Technology 2004, 64, 299-308.

[7] Tosun Gul and Muratoglu Mehtap. *The drilling of Al/SiCp metal-matrix composites. Part II: work piece surface integrity',* Composite Science and Technology 2004, 64, 1413-1418.

[8] Ramulu M., Rao P.N. and Kao H. *Drilling of (Al2O3)p/6061 metal matrix composites,* Journal of Material Processing Technology 2002, 37, 373-389.

[9] Davim J. Paulo and Baptista A. Monteiro. *Cutting force, tool wear and surface finish in drilling of metal matrix composites, IMechE* - Journal of Process Mechanical Engineering part E 2001, 215, 177-183.

[10] Mubaraki B., Bandyopadhyay S., Fowle R. and Mathew P. *Drilling studies of an Al2O3-Al metal matrix composite Part I Drill Wear Characteristics,* Journal of Material Science 1995, 30, 6273-6280.

[11] Monaghan J. and O'Reilly P. (1992) *'The drilling of an Al/SiC metal matrix composite',* Journal Material Processing Technology,' 33, 469-480.

[12] Sangkee Min, David A Dornfeld and Yohichi Nakao. *Influence of exit surface angle on drilling burr formation,* Journal of Manufacturing science and Engineering 2003, 125, 637-644.

[13] Gillespie L.K. and Blotter P.T. *The Formation and Properties of Machining Burrs*, Journal of Engineering for Industry 1976, 98, 66-74.

[14] Iwata K., Ueda K. and Okuda K. *Study of mechanism of burrs formation in cutting based on direct SEM observation*, Journal of the Society of Precision Engineering 1982, 48-4, 510-515.

[15] David A. Dornfeld, Sung and Lim Ko. *A study on burr formation mechanism,* Journal Engineering Materials and Technology 1991, 113, 75-87.

[16] Guo Y. B. Dornfeld D.A. and Nov. *Finite element modeling of burr formation in drilling 304 stainless steel*, Journal of Manufacturing science and Engineering 2000, 122, 612-619.

[17] Barnes Stuart and Pashby I.R. Through tool Coolant *Drilling of Aluminum/SiC Metal Matrix Composite*, Journal of Engineering Materials and Technology 2000, 122, 384-388.

[18] Sharma S.C., Girish B.M., Kulkarni R.S. and Kamath R. *Drillability of zinc/graphite metal matrix composites,* NML Technical Journal 1996, 38, 127-131.

[19] Songmene V. and Balazinski M. *Machinability of graphitic metal matrix composites as a function of reinforcing particles*, Annals of CIRP 1999, 48, 77-80.

[20] Songmene V. Balout B. and Masounave J. *Drilling of metal matrix Composites: cutting forces and chip formation', Light Metals (Metaux Legers)* International Symposium Light Metals as held at the 41st Annual Conference of Metallurgists of CIM (COM 2002); Montreal, Quebec; Canada, 11-14 Aug 2002, 633-651.

[21] S. Basavarajappa. *Investigations on dry sliding wear behaviour and machining characteristics of hybrid metal matrix composites*, Ph.D Thesis, Anna University,Chennai, 2006.

[22] Grum J. and Kisin M. Influence of microstructure on surface integrity in turning-part I: *The influence of the size of the soft phase in a microstructure on surface-roughness formation,* International Journal of Machine Tools and Manufacture 2003, 43, 1535-1543.

[23] Navarro P. Nathan. *Improved Surface Integrity with Borazon abrasive*, Society of Manufacturing Engineers 1971, Paper No. MR-71-803.

[24] Khales F. John and Micheal Field. *Surface integrity Guidelines for machining,* Society of Mechanical Engineers 1971, IQ 71-240.

[25] Novovic D., Dewes R.C., Aspinwall D.K., Voice W. and Bowen P. *The Effect of machined topography and integrity on fatigue life*, International Journal of Machine Tools and Manufacture 2003, 44, 125-134.

[26] Tosun Gul and Muratoglu Mehtap. *The drilling of Al/SiCp metal-matrix composites. Part II: work piece surface integrity*, Composite Science and Technology 2004, 64, 1413-1418.

[27] El-Gallab and M. Sklad. *Machining of Al/SiC particulate metal matrix composites part II: work piece surface integrity*, Journal of Materials Processing Technology 1998, 83, 277-285.

[28] S. Basavarajappa, G. Chandramohan, M. Prabhu, K. Mukund, M. Ashwin. *Drilling of Hybrid Metal Matrix Composites Work Piece Surface Integrity*, Int. Journal of Machine Tools and Manufacture, 2007, 47, 92-96.

[29] Davim J. Paulo and Antonio Conceicao C.A. *Optimal drilling of particulate metal matrix composites based on experimental and numerical procedures,* International Journal of Machine Tools and Manufacturing 2001, 41, 21-31.

[30] Davim J. Paulo. *Study of drilling metal-matrix composites based on the Taguchi Techniques,* Journal of Material Processing Technology 2003, 132, 250-254.

[31] Ramulu M., Kim D. and Kao H. *Experimental study of PCD tool performance in drilling (Al2O3)/6061 Metal Matrix composites*, Society of Manufacturing Engineers 2003, Paper No.MR 03-171.

In: Metal Matrix Composites
Editor: J. Paulo Davim

ISBN: 978-1-61209-771-8

Chapter 8

OPTIMIZATION OF MACHINING PARAMETERS FOR MULTIPLE PERFORMANCES IN DRILLING HYBRID COMPOSITES USING DESIRABILITY-BASED APPROACH

K. Palanikumar[1*], T. Rajmohan[2] and J. Paulo Davim[3]
[1]Sri Sairam Institute of Technology, Chennai, India
[2]Sri Chandrasekharendra Saraswathi Viswa Maha Vidyalaya University, Kanchipuram, India
[3]Department of Mechanical Engineering, University of Aveiro, Campus Santiago, Aveiro, Portugal

ABSTRACT

This chapter focuses on the multiple performances analysis in machining characteristics of drilling Hybrid metal matrix Composites produced through stir casting route. Desirability function based approach is employed for the optimization of drilling parameters namely spindle speed, feed rate and wt % of SiC based on the multiple performance characteristics including thrust force, surface roughness, and torque Experiments are conducted on Al 356- aluminum alloy reinforced with silicon carbide particles of size 25 microns and Mica of size 45 microns. Drilling test is carried out using coated HSS drill of 6 mm diameter. According to the Taguchi's quality concept, a L_{27}, 3-level orthogonal array was chosen for the experiments. An empirical model has been developed for predicting the thrust force, surface roughness and torque in drilling of Al356/SiC-Mica composites. The influences of different parameters and their interactions are studied in detail and presented in this study.

[*]Corresponding author, E-Mail: palanikumar_k@yahoo.com

1. INTRODUCTION

Metal matrix composites are materials which combine a tough metallic matrix with hard ceramic reinforcement. The inclusion of an additional reinforcement phase makes them hybrid composites. Metal matrix composites, in general posses certain superior properties like low density, high specific, stiffness and strength, controlled coefficient of thermal expansion superior wear resistance and corrosion resistance. But their poor machinability attributed to the presence of ceramic reinforcements. Resulting in very high rates of tool wear restricts their wider spread.

Cutting speed, feed rate, drill, work piece material, drill point angle and coolant conditions are the drilling parameters which highly affect the performance measures. In order to improve machining efficiency, reduce the machining cost, and improve the quality of machined parts, it is necessary to select the most appropriate machining conditions. The setting of drilling parameters relies strongly on the experience of operators. It is difficult to utilize the highest performance of a machine because there are too many adjustable machining parameters. In order to minimize these machining problems, there is need to develop scientific methods to select cutting conditions for damage-free drilling of materials

The effect of the various cutting parameters on the surface quality and microstructure on drilling of Al/17% SiC particulate MMC by using various drills was investigated by Tosun and Mehtap Muratoglu (2004). They have suggested that TiN coated HSS drills can be used for drilling Al/SiC-MMC rather than solid carbide tools. The study of drilling metal matrix composites of type A356/20% SiC-T6 based on the Taguchi technique with the objective of establishing the correlations between cutting velocity, feed rate and cutting time with the evaluation of tool wear, the specific cutting pressure and the hole surface roughness using PCD drill was investigated by Davim (2003)..

It is necessary to optimize the cutting parameters to obtain an extended tool life and better productivity, which are influenced by cutting force and torque (Shew and Kwong, 2002). Coated, uncoated high speed steel, carbide, PCD-tipped drills and solid carbide drills were used in the drilling tests. The results indicate that the hardness of the tool material has a significant influence on cutting edge wear and on the drilling-torque, surface finish and thrust-forces. (Monaghan, Reilly,1992).

Taguchi technique is a powerful tool for the design of high quality systems. It provides a simple, efficient and systematic approach to optimize design for performance, quality and cost. The methodology is valuable when design parameters are qualitative and discrete. Taguchi parameter design can optimize the performance characteristics through the setting of design parameters and reduce the sensitivity of the system performance to the source of variation (Ross, 1988; Taguchi, .and Konishi, 1987). In order to get good surface quality and dimensional properties, it is necessary to employ optimization techniques to find optimal cutting parameters and theoretical models to do predictions. Taguchi and response surface methodologies can be conveniently used for these purposes. The response surface method and genetic algorithm were used for predicting the surface roughness and optimizing the process parameters (Suresh et.al, 2002). Taguchi and response surface methodologies have applied for optimizing geometric errors in surface grinding process (Kwak, 2005).

Desirability functions have been used extensively to simultaneously optimize several responses. Since the original formulation of these functions contains non –differentiable points, only search methods can be used to optimize the overall desirability response. Furthermore all responses are treated as equally important (Harrington, 1965). Desirability function based approach was used for the optimization of drilling parameters for minimizing the delamination factor at entry and exit in drilling of MDF boards by Prakash et al, (2009).

For prediction, the response-surface method (RSM) is used and is practical, economical, and relatively easy to use. Response-surface methodology was used for the prediction of surface roughness in machining of GFRP composites (Palanikumar, 2008) and also for predicting performance in CNC drilling (Godfrey and Shivendra Kumar, 2006). The details of the mechanics of response-surface methodology are presented elsewhere. Further, the optimization of the process parameters is the important one and it improves productivity of the process. (Cochran and Cox, 1992).

In the present work, experiments are conducted on CNC machining center for drilling of Al356/SiC-Mica composites. Taguchi's L27 orthogonal array with three factors is considered for experimentation. Empirical relations were developed to predict the thrust force, surface roughness and torque using response-surface methodology. Analysis of variance (ANOVA) is used for checking the validity of the developed model. The multiple performance optimization of drilling parameters is carried out using response-surface methodology based on desirability function approach. The results indicate that the developed model can be effectively used to predict the responses in drilling of Al 356/SiC-Mica composites. The optimized results are validated using validation experiments. The influence of parameters and their interactions are studied in detail.

2. EXPERIMENTAL

2.1 Materials Used

Aluminum alloy Al356 was used as a matrix material and its chemical composition was as shown in Table 1. The silicon carbide particles of size 25 microns and mica, an average size of 45 microns were used as the reinforcement materials. The composites were fabricated with 5-15 weight % of the SiC particle in steps of the 5 weight % and a fixed quantity of 3 weight % of Mica.

The composites were fabricated by stir casting method, which was used by the other researchers (Davim, 2003). The Al 356 alloy, which was in the form of ingot, was cut into small pieces to accommodate into the Silica Crucible. The mica for the study has been procured from Premier Mica Company Chennai, India. Aluminum alloy was first melted in an electric furnace. Mica and SiC, preheated to a temperature of about 620° C, were added to the molten metal at 750° C and stirred continuously. The stirring was done at 500 rpm for 5-7 min. Magnesium was added in small amounts during stirring to increase the wetting. The melt with reinforcement was poured into permanent metallic mould.

Table 1. Composition of Al 356 alloy

Parameter	Copper	Silicon	Magnesium	Manganese	Iron	Titanium	Zinc	Al
%	<0.0005	7.27	0.45	<0.002	0.123	0.08	0.005	Remaining

Table 2. L27 Orthogonal array used for experimentation

	Column numbers												
	V	f	$V x f$	$V x f$	R	$V x R$	$V x R$	$f x R$	-	-	$f x R$	-	-
Trial No.	1	2	3	4	5	6	7	8	9	10	11	12	13
1	1	1	1	1	1	1	1	1	1	1	1	1	1
2	1	1	1	1	2	2	2	2	2	2	2	2	2
3	1	1	1	1	3	3	3	3	3	3	3	3	3
4	1	2	2	2	1	1	1	2	2	2	3	2	3
5	1	2	2	2	2	2	2	3	3	3	1	3	1
6	1	2	2	2	3	3	3	1	1	1	2	1	2
7	1	3	3	3	1	1	1	3	3	3	2	3	2
8	1	3	3	3	2	2	2	1	1	1	3	1	3
9	1	3	3	3	3	3	3	2	2	2	1	2	1
10	2	1	2	3	1	2	3	1	2	3	1	3	1

11	2	1	2	3	2	3	1	2	3	1	2	1	2
12	2	1	2	3	3	1	2	3	1	2	3	2	3
13	2	2	3	1	1	2	3	2	3	1	3	1	3
14	2	2	3	1	2	3	1	3	1	2	1	2	1
15	2	2	3	1	3	1	2	1	2	3	2	3	2
16	2	3	1	2	1	2	3	3	1	2	2	2	2
17	2	3	1	2	2	3	1	1	2	3	3	3	3
18	2	3	1	2	3	1	2	2	3	1	1	1	1
19	3	1	3	2	1	3	2	1	3	2	1	2	1
20	3	1	3	2	2	1	3	2	1	3	2	3	2
21	3	1	3	2	3	2	1	3	2	1	3	1	3
22	3	2	1	3	1	3	2	2	1	3	3	3	3
23	3	2	1	3	2	1	3	3	2	1	1	1	1
24	3	2	1	3	3	2	1	1	3	2	2	2	2
25	3	3	2	1	1	3	2	3	2	1	2	1	2
26	3	3	2	1	2	1	3	1	3	2	3	2	3
27	3	3	2	1	3	2	1	2	1	3	1	3	1

Table 3. Machining parameter and levels

Control parameters	symbol	Level		
		1	2	3
Speed rpm	*V*	1000	2000	3000
Feed rate mm/min	*f*	50	100	150
Wt fraction of Sic %	*R*	5	10	15

2.2 Experimental Design

The experiments are conducted as per the standard orthogonal array. The selection of the orthogonal array is based on the condition that the degrees of freedom for the orthogonal array should be greater than or at least equal sum to those of machining parameters. In the present investigation an L_{27} orthogonal array was chosen, which has 27 rows and 13 columns as shown in Table 2. The machining parameter chosen for the experiment are (i) spindle speed (ii) feed rate (iii) wt % Sic. Table3 indicates the factors and their level. The experiment consists of 27 tests (each row in the L_{27} orthogonal array) and the columns were assigned with parameters. The first column was assigned to Speed (*V*), second column was assigned to feed rate (*f*), fifth column was assigned to wt % SiC (*R*) and the remaining columns were assigned to their interactions.

2.3 Experimental Procedure

Drilling tests are conducted on vertical CNC machining center. The machining samples were prepared in the form of 150mm×150mm×10mm blocks. The coated HSS drill bits of 6mm diameter were used. The surface roughness of the work piece was measured with a Mitutoyo portable Surftest SJ-201 P/M contact profilometer. Direction of surface roughness measurements was perpendicular to the hole circumference. The surface roughness values given in this study are the mathematical average of two measurements taken from the same hole surface. The computer controlled data acquisition system was used to collect and record the data of the experiments. The Kirsler dynamometer was used to record the thrust force and torque. The dynamometer is connected to a 3-channel charge amplifier type through a connecting cable, which in turn is connected to the PC by a 37-pin cable from the A/D board. The experiments are repeated twice to circumvent the possible experimental errors A schematic arrangement of experimental setup is shown in Figure 1. The sample outputs of the dynamometer (variation of thrust force, Torque with acquisition time) at different spindle speeds are given in Figures 2-3. The observed response values for all the 27 experiments are listed in Table 4.

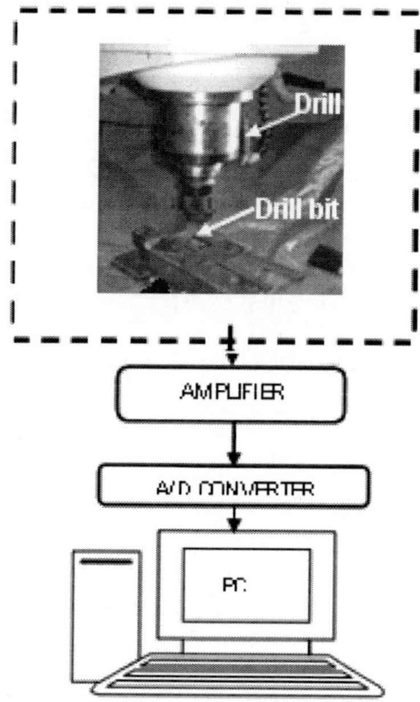

Figure 1. Schematic representation of experimental set-up.

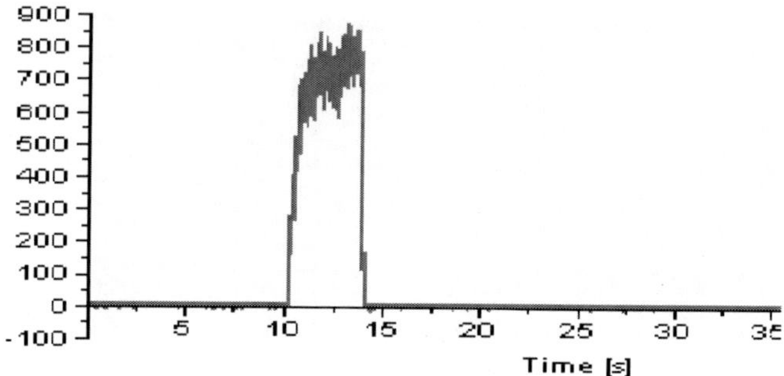

Figure 2. Typical thrust force observed when speed =1000 rpm; feed=150 mm/min and work piece contains 15% SiC +3%Mica.

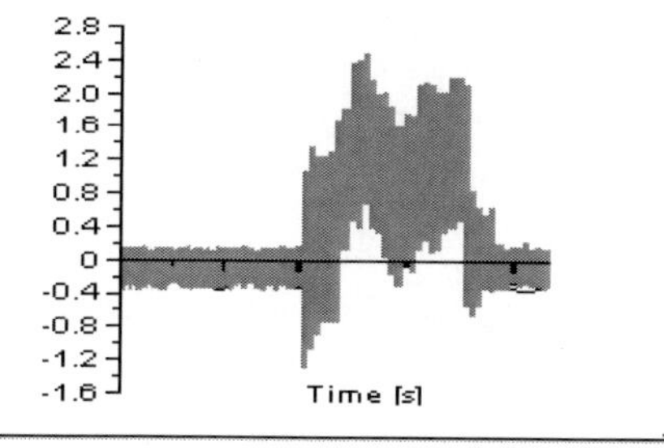

Figure 3. Typical Torque observed when speed =2000 rpm; feed=50 mm/min and work piece contains 5% SiC +3%Mica.

Table 4. Design of experiments based on L_{27} orthogonal array

Expt No.	Coded Values			Actual Setti			Thrust force	Roughness	Torque
	V	*f*	*R*	*V*	*f*	*R*	N	µm	Nm
1	1	1	1	1000	50	5	640	5.2	2.9
2	1	1	2	1000	50	10	675	5	3
3	1	1	3	1000	50	15	705	4.8	3.2
4	1	2	1	1000	100	5	735	5.5	3.3
5	1	2	2	1000	100	10	755	5.3	3.4
6	1	2	3	1000	100	15	795	5	3.7
7	1	3	1	1000	150	5	820	6	3.8
8	1	3	2	1000	150	10	845	5.8	3.85
9	1	3	3	1000	150	15	860	5.5	4
10	2	1	1	2000	50	5	640	4.2	2.95
11	2	1	2	2000	50	10	675	4	3
12	2	1	3	2000	50	15	705	3.75	3.2
13	2	2	1	2000	100	5	735	4.5	3.3
14	2	2	2	2000	100	10	755	4.2	3.6
15	2	2	3	2000	100	15	795	4	3.75
16	2	3	1	2000	150	5	820	5.1	3.8
17	2	3	2	2000	150	10	845	4.7	4
18	2	3	3	2000	150	15	860	4.4	4.1
19	3	1	1	3000	50	5	690	3.4	2.9
20	3	1	2	3000	50	10	710	3	3
21	3	1	3	3000	50	15	740	2.8	3.1
22	3	2	1	3000	100	5	780	3.6	3.5
23	3	2	2	3000	100	10	805	3.3	3.75
24	3	2	3	3000	100	15	855	3	3.9
25	3	3	1	3000	150	5	870	4.4	3.95
26	3	3	2	3000	150	10	905	4.1	4.1
27	3	3	3	3000	150	15	925	3.8	4.15

3. Modeling Drilling Parameters Using Response-Surface Methodology

Response-surface methodology (RSM) is one of the important techniques for determining the relationship between various process parameters with the various drilling criteria and exploring the effect of these process parameters on the coupled responses (Palanikumar, 2008). Response-surface method is used to establish the mathematical relation between the output response y and the various drilling parameters. The general second order polynomial response-surface model, used to evaluate the parametric effects on the drilling of Al 356/SiC-Mica is as follows

$$Y = \beta_0 + \sum_{i=1}^{k} \beta_i x_i + \sum_{i=1}^{k} \beta_{ii} x_i^{\,2} + \sum_i \sum_j \beta_{ij} x_i x_j + \varepsilon \qquad (1)$$

Where y is the response and x_i are the values of the drilling parameters. The term β are the regression coefficients. The second term under the summation sign of this response-surface equation is attributable to linear effect, whereas the third term corresponds to the higher-order effects; the fourth term of the equation includes the interactive effects of the process parameter. The residual ε measures the experimental error of the observations. The second order response surface representing the thrust force, roughness and torque can be expressed as a function of drilling parameters such as spindle speed (*V*), feed (*f*) and wt % of SiC (*R*). The relationship between the thrust force and machining parameters are expressed as follows:

Thrust force = +502.77778 + 6.66667 *WtofSiC +1.71111 *Feed rate +0.013333*Spindle speed - 0.015000 * Wt of SiC * Feed rate +2.50000E-004* Wt of SiC * Spindle speed +9.16667E-005* Feed rate * Spindle speed (2)

R-Sq =97.2%

The validity of the model developed is evaluated by using analysis of variance (ANOVA). Analysis of variance essentially consists of partitioning the total variation in an experiment into components ascribable to the controlled factors and error. Table 5 shows the results of ANOVA for thrust force. In ANOVA table, the sum of squares are used to estimate the square of deviation from the grand mean. Mean squares are estimated by dividing the sum of squares by degrees of freedom. F-ratio is an index used to check the adequacy of the model in which calculated value of F should be greater than the F table value. The Model F value of 115.95 implies that the model is significant. There is only a 0.01% chance that a "Model F-Value" this large could occur due to noise.

The adequacy of the model is further analyzed by using R-Sq values. The values of R-sq represent the regression coefficient. The larger value of R-Sq is always desirable (Montgomery, 1991). In the present case, the R-Sq value is 97.2% which shows the high correlation that exists between the experimental values and predicted values. Also, experimental and predicted data by using the aforesaid models are plotted as shown in Figs. 4 and 5, which indicate a good correlation between the model and experimental values and

Table 5. Analysis of variance of response-surface model for Thrust force

Source	Sum of Squares	df	Mean square	F value	P-value
Model	1.63E+05	6	27179.75	115.95	< 0.0001
V-Speed	14450	1	14450	61.64	< 0.0001
f-Feed	1.37E+05	1	1.37E+05	584.19	< 0.0001
R-Wt of SiC	11250	1	11250	47.99	< 0.0001
Vf	168.75	1	168.75	0.72	0.0996
VR	18.75	1	18.75	0.08	0.8930
fR	252.08	1	252.08	1.08	0.0306
Residual	4688.19	20	234.41		
Cor Total	1.68E+05	26			

hence, the developed model can be effectively used to predict the thrust force in drilling of Al 356/SiC-Mica within the limits of the factors studied. Similarly, the model is created for surface roughness and torque. The mathematical relationship established for correlating surface roughness, torque and machining parameters in drilling Al 356/SiC-Mica is given by

Surface roughness = +5.94444 - 0.028889 * Wt of SiC +7.00000E-003* Feed rate - 9.94444E-004 * Spindle speed -1.16667E-004* Wt of SiC * Feed rate -6.66667E-006 * Wt of SiC * Spindle speed +1.33333E-006* Feed rate * Spindle speed (3)

R-Sq =97.8%

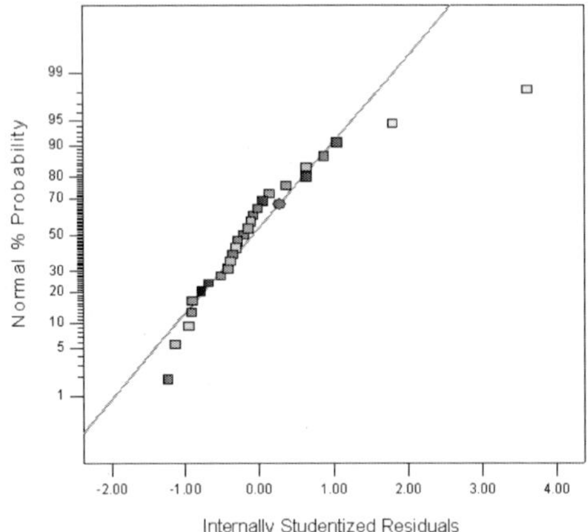

Figure 4. Normal probability plot for Thrust force.

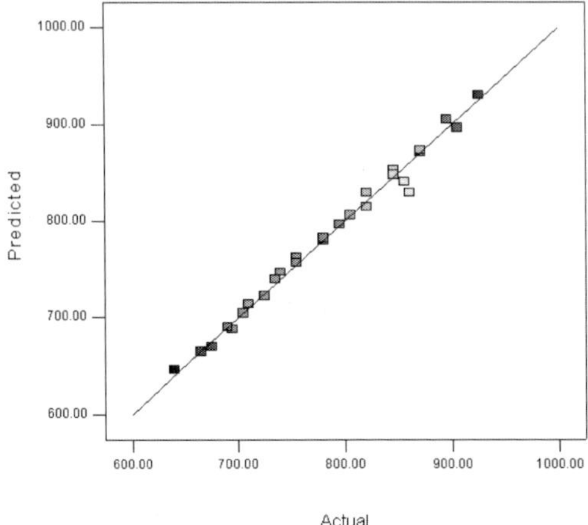

Figure 5. Correlation graph for thrust force.

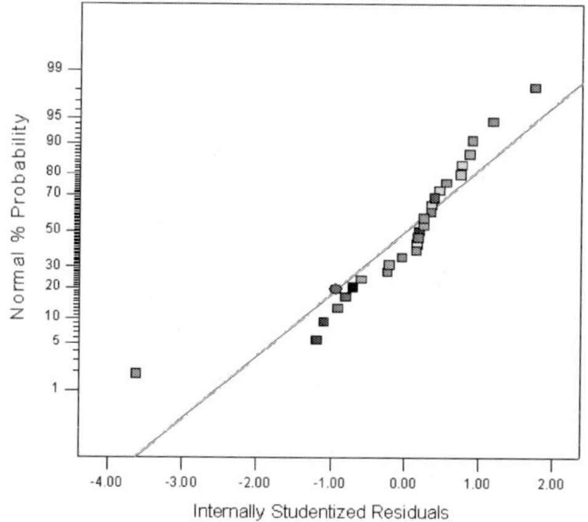

Figure 6. Normal probability plot for surface roughness.

Torque =+ 2.31481 + 0.035000* Wt of SiC+7.44444E-003 * Feed rate -2.50000E-005* Spindle speed-1.66667E-005 * Wt of SiC * Feed rate-1.66667E-006* Wt of SiC * Spindle speed +1.08333E-006 * Feed rate * Spindle speed (4)

R-Sq = 97.18%

The ANOVA table for surface roughness is presented in Table 6. The Model F value of 205.7 implies that the model is significant. There is only a 0.01% chance that a "Model F-Value" this large could occur due to noise. The values of R-sq represent the regression confidence. In the present case, the R-Sq value is 97.8%, which shows the high correlation that exists between the experimental values and predicted values.

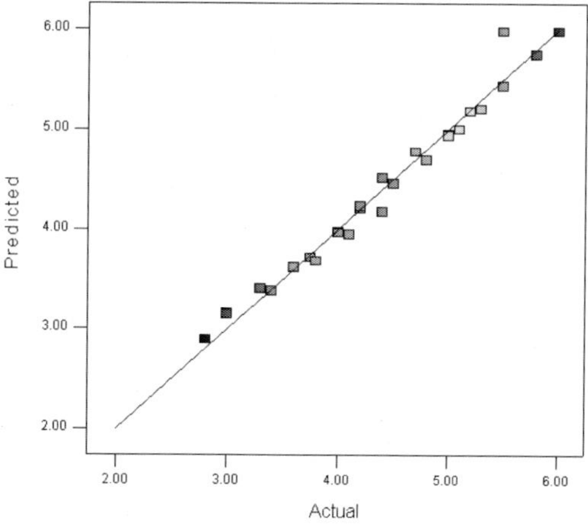

Figure 7. Correlation graph for surface roughness.

Also, experimental and predicted data by using the aforesaid models are plotted as shown in Figs(6-7)which indicate a good correlation between the model and experimental values and hence, the developed model can be effectively used to predict the surface roughness in drilling of Al 356/SiC-Mica within the limits of the factors studied.

The ANOVA table for Torque is presented in Table 7. The Model F value of 114.77 implies that the model is significant. There is only a 0.01% chance that a "Model F-Value" this large could occur due to noise. The values of R-sq represent the regression confidence. In the present case, the R-Sq value is 97.18%, which shows the high correlation that exists between the experimental values and predicted values. Also, experimental and predicted data by the using the aforesaid models are plotted as shown in Figs(8-9),which indicate a good correlation between the model and experimental values and hence, the developed model can be effectively used to predict the Torque in drilling of Al 356/SiC-Mica composites within the limits of the factors studied.

Table 6. Analysis of variance of response-surface model for Surface roughness

Source	Sum of Squares	df	Mean square	F value	P-value
Model	20.13	6	3.35	205.7	< 0.0001
V-Speed	1.31	1	1.31	80.13	< 0.0001
f-Feed	3.25	1	3.25	199.35	< 0.0001
R-Wt of SiC	15.49	1	15.49	950.02	< 0.0001
Vf	0.01	1	0.01	0.63	0.0016
VR	0.013	1	0.013	0.82	0.1691
FR	0.053	1	0.053	3.27	0.3640
Residual	0.33	20	0.016		
Cor Total	20.45	26			

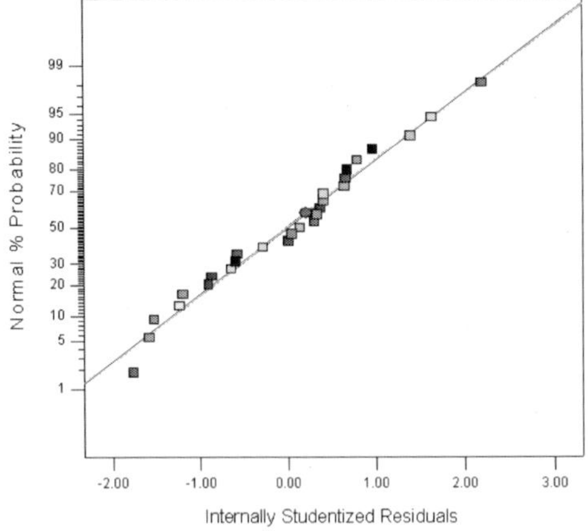

Figure 8. Normal probability plot for Torque.

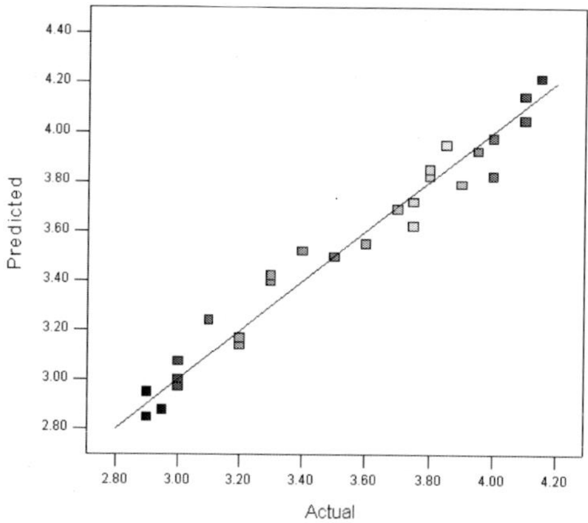

Figure 9. Correlation graph for Torque.

Table7. Analysis of variance of response-surface model for Torque

Source	Sum of Squares	df	Mean square	F value	P-value
Model	4.54	6	0.76	114.77	< 0.0001
V-SPeed	0.41	1	0.41	61.5	0.0002
f-Feed	4.01	1	4.01	609.49	< 0.0001
R-Wt of SiC	0.08	1	0.08	12.15	< 0.0001
Vf	2.08E-04	1	2.08E-04	0.032	0.0011
VR	8.33E-04	1	8.33E-04	0.13	0.0929
FR	0.035	1	0.035	5.35	0.0107
Residual	0.13	20	6.59E-03		
Cor Total	4.67	26			

4. Optimization of Drilling Parameters Using Response-Surface Methodology-Based Desirability Function Approach

Optimization of process parameters is important and it improves productivity. Desirability based optimization technique has produced a unique and powerful optimization procedure that differs from traditional practices. In recent years, desirability function approach is used by some of the researchers for finding the optimal solutions using multi-performance objective. (Prakash et.al.,2009; Shew and Kwong, 2002;)

Desirability is an objective function that ranges from zero outside of the limits to one at the goal. The numerical optimization finds a point that maximizes the desirability function. The characteristics of a goal may be altered by adjusting the weight or importance. For several responses and factors, all goals get combined into one desirability function. Desirability was described using multi responses (Montgomery, 1991). The method makes use of an objective function, *D(x)*, called the desirability function. It reflects the desirable ranges for each response (d_i). The desirable ranges are from zero to one (least to most desirable, respectively). The simultaneous objective function is a geometric mean of all transformed responses:

$$D = (d_1 \text{ x } d_2 \text{ x.......x } d_n)^{1/n} = \left(\prod_{i=1}^{n} d_i\right)^{\frac{1}{n}} \tag{5}$$

where n is the number of responses in the measure. If any of the responses or factors fall outside their desirability range, the overall function becomes zero.

In the present work, the response-surface methodology-based desirability approach is used for the optimization of drilling parameters namely spindle speed, feed and wt % of SiC based on the multiple performance characteristics including thrust force, surface roughness, and torque. In desirability function approach, the measured properties of each predicted response is transformed to a dimensionless desirability values. The scale of desirability function ranges between d=0, which suggests that the response is completely unacceptable, and d=1, which suggests that the response is exactly the target value. The value of d increases as the desirability of the corresponding response increases (Shew and Kwong, 2002; Gaitonde et.al, 2008). In desirability-based approach, one-sided transformation is used to transform the response into a desirability value. In this study the transformation of characteristics are smaller-the-better characteristic. A weight can be assigned to a goal to adjust the shape of its particular desirability function. For a goal of minimum the desirability will be defined by the following formula and desirability curves for goal minimum is presented in Fig. 10.

$$d_i = 1 \qquad \mathrm{Y_i} \leq Low_1 \mathrm{Y} \tag{6}$$

$$d_i = \left[\frac{\mathrm{High_i} - Y_i}{High_i - Low_i}\right]^{\mathrm{Wt_i}}, \mathrm{Low_i} < \mathrm{Y_i} < \mathrm{High_i} \tag{7}$$

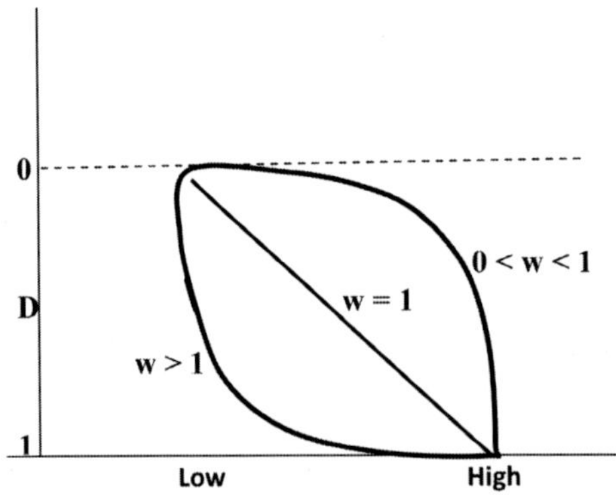

Figure 10. Desirability curves for goal minimum.

Table 8. Goals set and limits used for the optimization

Parameter and response	Goal	Lower limit	Upper limit	Lower weight	Upper weight	Importance
Spindle speed	Is in range	1000	3000	1	1	3
Feed rate	Is in range	50	150	1	1	3
Wt % SiC	Is in range	5	15	1	1	3
Thrust force	Minimize	640	925	1	1	3
Surface roughness	Minimize	3	5.8	1	1	3
Torque	Minimize	2.9	4.15	1	1	3

Table 9. Best global solution for optimization

No	Speed	Feed Rate	Wt % SiC	Thrust force	Roughness	Torque	Desirability
1	3000	50	15	675	3.23	2.92	0.906
2	2999	50	14.98	675.7	3.33	2.95	0.906
3	2999	50	14.97	675.6	3.43	2.96	0.905

Table 10. Optimum Values

Parameters	Values
Spindle Speed	3000 rpm
Feed rate	50 mm/min
Wt % SiC	15

$$d_i = 0 \qquad \mathrm{Y_i} \geq \mathrm{High_i} \tag{8}$$

The optimization analysis was carried out using DESIGN-EXPERT® software. The goal set, lower limits used, upper limits used, weights used, and importance of the factors are presented in Table 8. In desirability-based approach, different best solutions are obtained. The solution with high desirability is preferred. The best three solutions obtained for the surface roughness optimization is presented in Table 9. From the table, the global solution obtained for minimization of thrust force, surface roughness and torque in drilling of Hybrid Al 356/Sic/Mica MMC by solid carbide drills is presented in Table 10.

5. ANALYSES OF EXPERIMENTAL RESULTS AND DISCUSSIONS

Hybrid composite materials are important materials, and are finding increased applications in many engineering components. These composites have abrasive particles. These abrasive particles pose big problems during machining. Spindle speed, federate and wt % of SiC are the major drilling parameters that are considered in the experiments. The thrust force, torque developed and surface roughness in drilling operation is an important concern. Monitoring of thrust force, torque and surface roughness in drilling is needed for the industry. For reducing of thrust force, surface roughness and torque, optimization of process parameters are required. In this work the multiple performance optimization of drilling parameters is carried out using response-surface methodology based on desirability function approach.

The optimization is carried out for combination of goals. The goals apply to factors and responses. The goals used for thrust force, torque and surface roughness are "minimize" and the goals used for the factors are "within range". A weight can be assigned to a goal to adjust the shape of its particular desirability function. Figure 11 shows the histogram of desirability for the three best solutions. This is a simple view that shows the desirability for each factor and response individually. It can be generated for each optimum found. The solutions are sorted with the most desirable first. By default, the input factors are set range, thus preventing extrapolation. In the figure, the first three bars show the input parameters. The next three bars shows the desirability obtained for Thrust force, torque and surface roughness. Finally, the seventh bar indicates the combined desirability.

The optimization results indicated that high spindle speed, low feed rate, and high wt % of SiC are preferred for drilling of hybrid Al356/SiC Mica MMC. The increase in spindle speed reduces the surface roughness in drilling of hybrid Al356/SiC Mica MMC. The reason being at high spindle speed, the force induced in cutting has been increased which in turn cut the hybrid MMCs smoothly and produces lesser surface roughness than slow speed. It can also be observed that there is no predominant variation in thrust force and torque on increasing the spindle speed for all the feederates The results indicated that low feed rate is preferred in drilling of hybrid Al356/SiC Mica MMC. The increase in feed rate increases the thrust force and torque and it leads to poor surface finish. Also, the increase in feed rate increases the heat generation during drilling which in increases the surface roughness. The increase in wt % of SiC decreases the surface roughness.

The experimental results of thrust force, surface roughness and torque as a function of cutting parameters in drilling of composites as per orthogonal array are recorded and analysed. ANOVA Table (Tables 5-7) clearly shows that feed rate and Speed are the two dominant parameters and hence contributes in improving surface quality, reduces the thrust force and torque.

Estimated contour plots for Desirability, Thrust force, surface roughness and torque on Speed Vs Feed planes was presented in Figs. 12-15. The figure shows that the increase of spindle speed increases the desirability and vice versa.

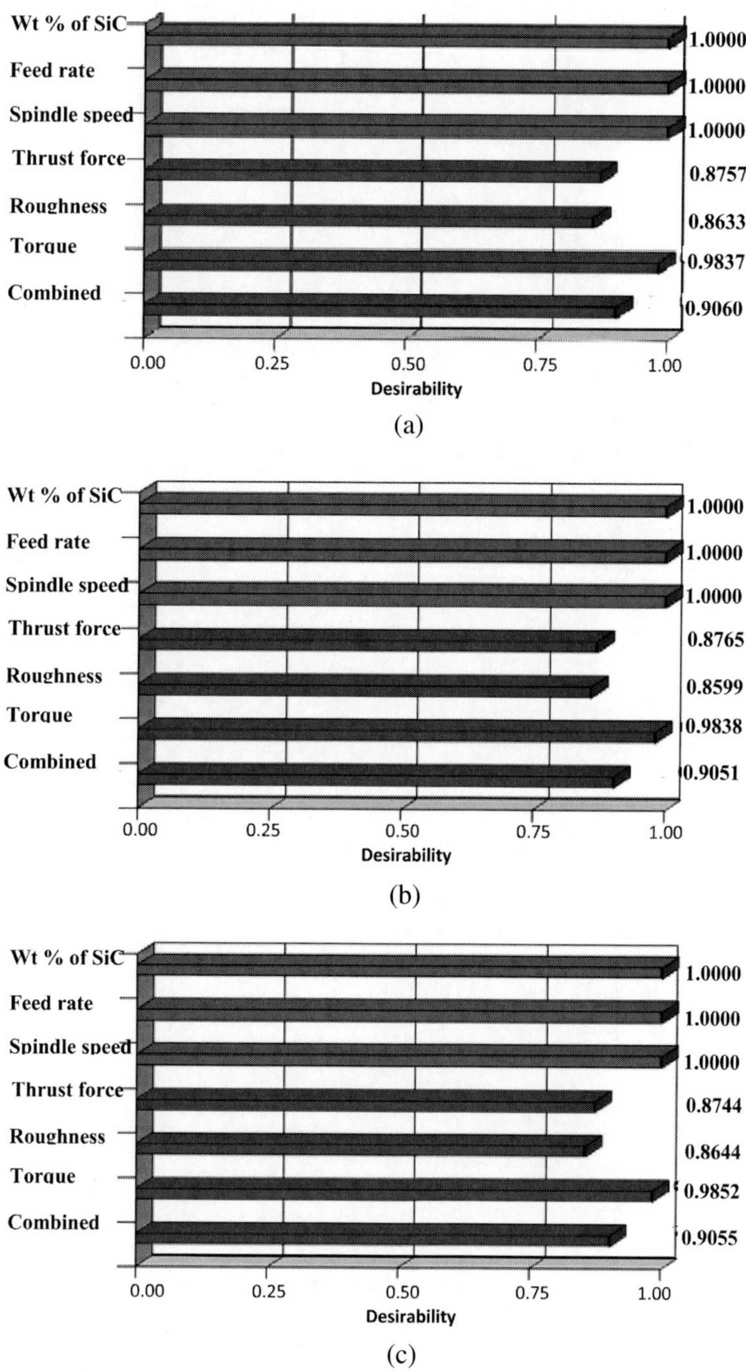

Figure 11. Histogram of best three solutions.

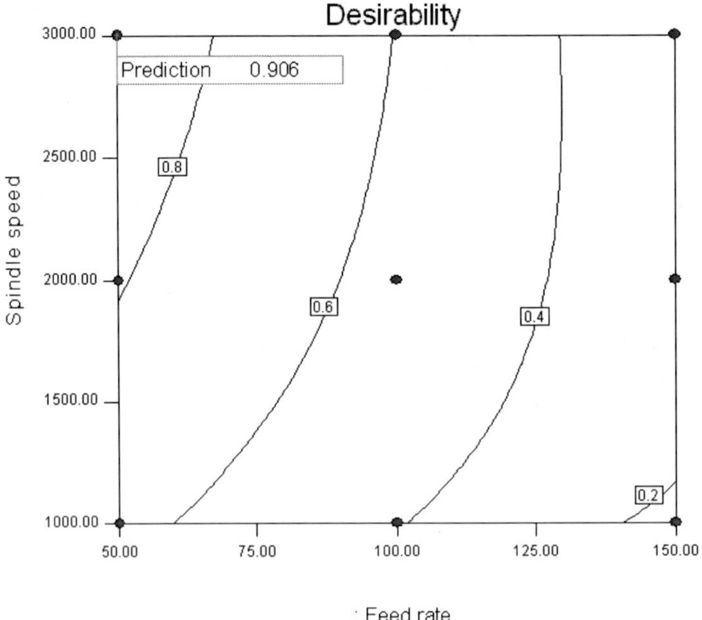

Figure 12. Estimated contour plots for Desirability (Wt % of SiC=15).

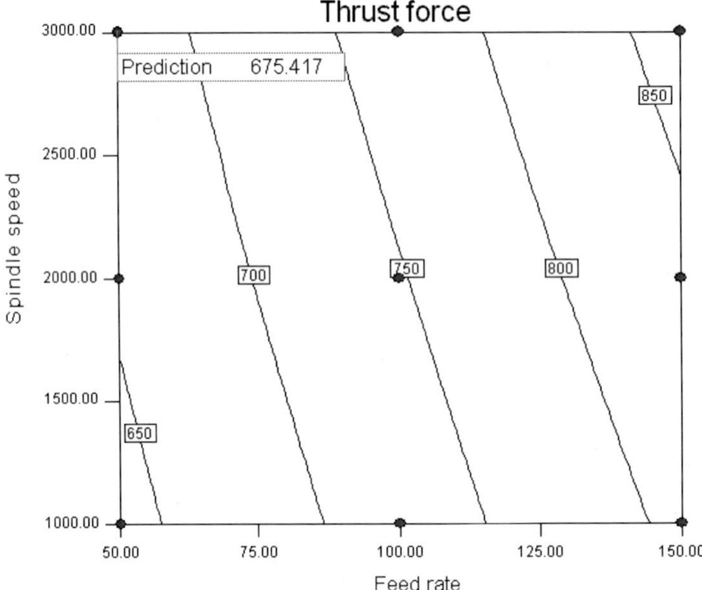

Figure 13. Estimated contour plots for Thrust force (Wt % of Sic=15).

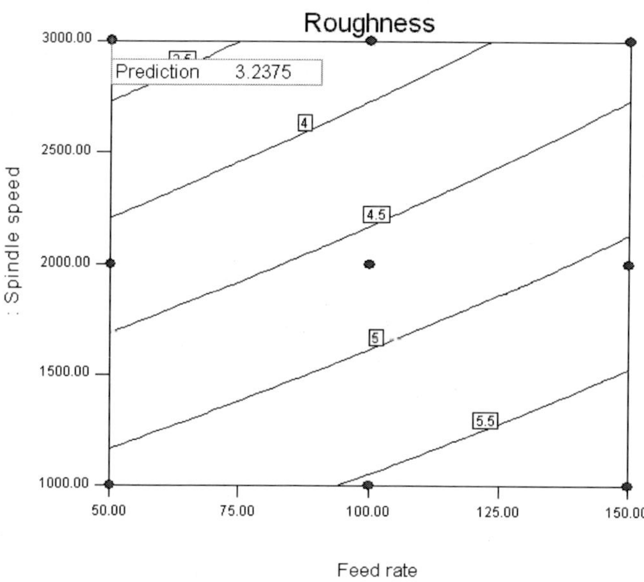

Figure 14. Estimated contour plots for Roughness (Wt % of Sic=15).

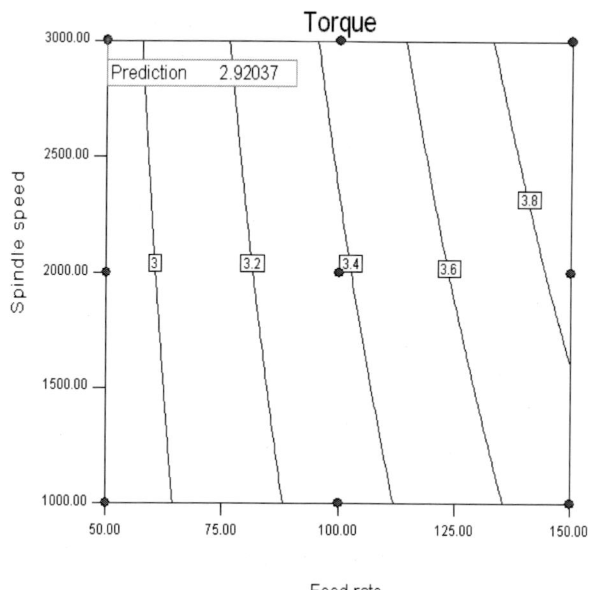

Figure 15. Estimated contour plots for Torque (Wt % of Sic=15).

The interaction effect of spindle speed Vs feed rate on responses in drilling of hybrid Al356/SiC Mica composites was shown in Figs. 16-19. The figure shows that the increase of feed rate decreases the desirability and vice versa.

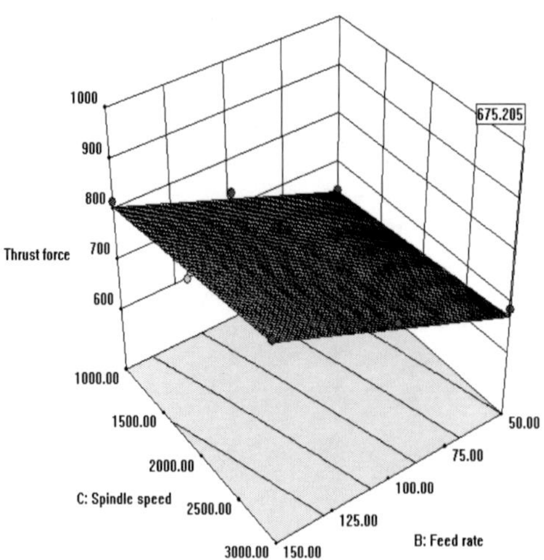

Figure 16. Estimated 3D response surface plot for thrust force at Wt % of SiC=15.

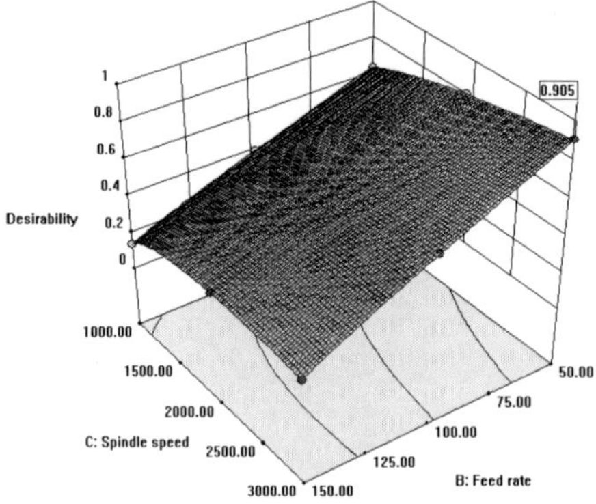

Figure 17. Estimated 3D response surface plot for Desirability at Wt % of SiC=15.

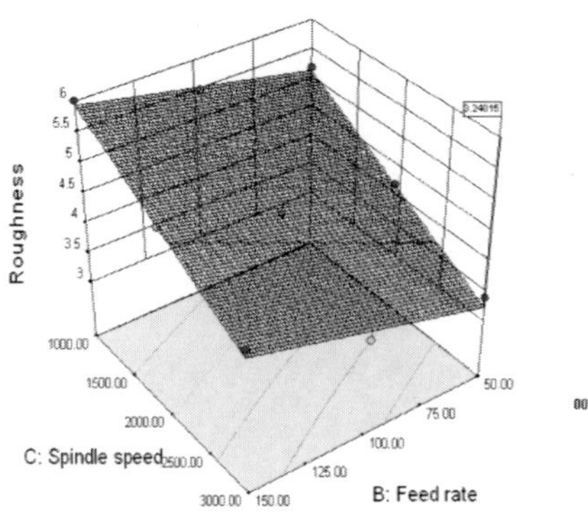

Figure 18. Estimated 3D response surface plot for Roughness at Wt % of SiC=15.

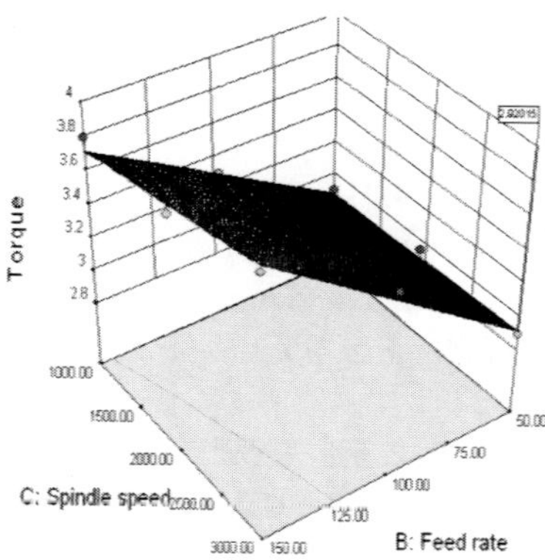

Figure 19. Estimated 3D response surface plot for Torque at Wt % of SiC=15.

Conclusion

In this chapter, the optimal drilling parameters were determined for the multi-performance characteristics (Thrust force, Torque and surface roughness) by using Response-surface modeling and desirability-based optimization. Based on the results, the following conclusions are arrived

1) Response-surface methodology is used for the modeling of drilling parameters in machining of Al 356/SiC-Mica composites. The results indicated that the models are effective to predict the responses in drilling composites.
2) Multiple response optimization is carried out using desirability-based approach. The optimal solution really reduces the responses in drilling of composites. The optimal solution obtained is near optimal and can be improved further by considering more variables and levels.
3) The results indicated that feed rate is the main parameter, which influences the thrust force, torque and roughness in drilling of composites.
4) The interactions between the parameters also have some effect on the drilling of hybrid MMC composites.
5) This technique will be more convenient and economical to predict the effects of different influential combinations of parameters within the levels studied.

References

Cochran WG, Cox GM (1992) *Experimental designs*. Wiley, NewYork

Davim JP (2003) *Study of drilling metal-matrix composites based on the Taguchi Techniques*. J Mater Process Technol 132:250–254

Godfrey CO, Shivendra Kumar (2006) *Response surface methodology-based approach to CNC drilling operations*. J Mater Process Technol 171:41–47

Harrington,E.C.,JR.(1965) *"The Desirability Function"*. Industrial quality control 21, pp494-498

Kwak J–S (2005) *Application of Taguchi and response surface methodologies for geometric error in surface grinding process*. Int J Mach Tool Manuf 45:327–334

Monaghan J, O_Reilly P. *The drilling of an Al/SiC metal–matrix composite*. J Mater Process Technol 1992;33:469–80.

Montgomery DC (1991) *Design and analysis of experiments*.Wiley, New York

Palanikumar K (2008) *Application of Taguchi and response surface methodologies for surface roughness in machining glass fiber reinforced plastics by PCD tooling*. Int J Advan Manuf Technol 36:19–27

Prakash.S, K. Palanikumar & N. Manoharan (2009) *Optimization of delamination factor in drilling medium-density fiberboards (MDF) using desirability-based approach*. Int J Advan Manuf Technol DOI 10.1007/s00170-009-1974-2

Ross, P.J., 1988. *Taguchi Technique for Quality Engineering*. McGraw-Hill, New York

Shew YW, Kwong CK (2002) *Optimization of the plated through hole(PTH) process using experimental design and response methodology*. Int J Adv Manuf Technol 20:758–764

Suresh PVS, Venkatehwara rao K, Desmukh SG (2002) *A genetic algorithmic approach for optimization of the surface roughness prediction model*. Int J Mach Tools Manuf 42: 675–680

Taguchi, G., Konishi, S., 1987. *Taguchi methods, orthogonal arrays and linear graphs. In: Tools for Quality Engineering*. American Supplier Institute, pp. 35–38.

Tosun G, Mehtap Muratoglu (2004) *The drilling of Al/SiCp metal matrix composites*. Part I: Microstructure, Compos Sci Tech 64:209–308

Tosun G, Mehtap Muratoglu (2004) *The drilling of Al/SiCp metal matrix composites*. Part II: Work piece Surface integrity, Compos Sci Tech 64:1413–1418

INDEX

D

E

F

M

N

O

P

R

S

T

V

W

X

Y

Z